The Rhubarb Connection and Other Revelations

The Everyday World of Metal Ions

The Rhubarb Connection and Other Revelations
The Everyday World of Metal Ions

By

Lars Öhrström

and

Jacques Covès

ROYAL SOCIETY
OF **CHEMISTRY**

Print ISBN: 978-1-78801-094-8
EPUB ISBN: 978-1-78801-669-8

A catalogue record for this book is available from the British Library

The Royal Society of Chemistry is a charity, registered in England and Wales, Number 207890, and a company incorporated in England by Royal Charter (Registered No. RC000524), registered office: Burlington House, Piccadilly, London W1J 0BA, UK, Telephone: +44 (0) 20 7437 8656.

Visit our website at www.rsc.org/books

Printed in the United Kingdom by CPI Group (UK) Ltd, Croydon, CR0 4YY, UK

Foreword

Our cases were separated by thousands of miles and over 100 years, but it turned out we were tracking the same culprit. No, not some villainous time traveler, like a warped Doctor Who. We were after an even *more* famous suspect who was, for a long time, even harder to track down. Number 33 of Group 15; Arsenic.

As a forensic chemist, I often chase and write crime stories. As an inorganic chemist, Lars has a special affinity for metals and metal-like elements. I was looking into a supposedly accidental poisoning of Roosevelt's Tree Army in America in 1937 for my *Chemistry World* column "Trace Analysis", while Lars delved into an 1829 English murder-by-poison case, the one that inspired James Marsh to develop his critical test for arsenic, for his wonderful book *The Last Alchemist in Paris*. We often find ourselves trading stories over the periodic table and the compounds its elements make up. Chemistry as a weapon or medicine, as a tool of vengeance or justice, as beautiful or horrible – and everything in between.

The complexity of chemistry is always expertly presented by Lars, as seen in *The Last Alchemist in Paris* and here in *The Rhubarb Connection*, where he has also collaborated with biochemist Jacques Covès. Lars and Jacques find page-turning ways to talk about chemistry. I initially focused on Chapter 4 of this book – it has "poisoning" in its title, after all – but soon I was caught. I just

The Rhubarb Connection and Other Revelations: The Everyday World of Metal Ions
By Lars Öhrström and Jacques Covès
© Lars Öhrström and Jacques Covès 2019
Published by the Royal Society of Chemistry, www.rsc.org

had to learn more about the connection between rubber and the anticancer drug Taxol®, the role of Pan Am airlines in drug trafficking, and what a harmless plant's acidic taste had to do with Napoleon's death. *The Rhubarb Connection* is a series of fantastic stories that showcase the central role metals play in our lives and sometimes our deaths.

Metals shine in *The Rhubarb Connection*, with their luster lighting up our recent history in fun and informative ways. In ways I plan to steal. As a chemistry professor and science communicator, I am always on the lookout for hooks that capture people's attention. *The Rhubarb Connection*, like *The Last Alchemist*, is full of great hooks. Like the ones that caught me and the ones that will grab you. So, yes, I confess that I will steal Lars' and Jacques' hooks. With proper citation, of course.

It will be a real crime that sees Lars and I cross paths again. Not ours, mind you! Another's foul deeds will bring us to some square on the periodic table – or perhaps our old spot in Period 4. In our own ways, we'll both continue tracking chemicals, culprits, and accounts. I look forward to our next chemistry meet-up because I know Lars will have a fascinating tale to share.

Raychelle Burks

About the Authors

Lars Öhrström is a Swedish inorganic chemist who has also worked in France, Botswana, Switzerland, and the USA. He was educated at the Royal Institute of Technology in Stockholm and currently he is professor at Chalmers University of Technology in Gothenburg. Lars is a regular contributor with popular texts to *Chemistry World* magazine, and *Nature Chemistry,* as well as to the Swedish *Sans* magazine. His main research interest is in Metal–Organic Frameworks, potentially porous materials with budding technological applications in green and sustainable chemical engineering. His critically acclaimed popular science book *The Last Alchemist in Paris* (2013) has been translated to four languages.

Jacques Covès is a research director at The French National Centre for Scientific Research, CNRS. He is a biochemist with a specific interest for metals in biology, and works in Grenoble, France at the Institut de Biologie Structurale. Jacques was responsible for the French edition of *The Last Alchemist in Paris*, (*Le dernier Alchimiste à Paris*, 2016) and has also translated university level textbooks to French.

The Rhubarb Connection and Other Revelations: The Everyday World of Metal Ions
By Lars Öhrström and Jacques Covès
© Lars Öhrström and Jacques Covès 2019
Published by the Royal Society of Chemistry, www.rsc.org

FOREWORD

Raychelle Burks is an assistant professor of chemistry at St. Edward's University, Austin, Texas. An in-demand speaker and host, Raychelle appears on the Science Channel's *Outrageous Acts of Science*, *ACS Reactions* videos, and at genre conventions such as *DragonCon* and *GeekGirlCon*. She writes a forensic science column for *Chemistry World*, the magazine of the Royal Society of Chemistry.

Acknowledgements

Our foremost thanks go to our families; Nina Kann, Agnes Öhrström Kann, Rebecka Öhrström Kann, Isabelle Michaud Soret, Lilou Covès, Marion Covès and Pablo Covès, who have once again coped with a book project and we are grateful for your support, discussions, encouragement and company while visiting odd museums and places near and far.

In addition to this, special thanks to Nina Kann and Isabelle Michaud Soret, our companions in life, who know some of the chemistry in this book far better than we, and sowed the seeds of inspiration for several chapters. We could not have made this without you.

We also thank our editor Michelle Carey and her co-workers at the Royal Society of Chemistry for believing in the project and keeping us on the straight and narrow (more or less).

Special thanks to Raychelle Burks for discussing analytical chemistry, gruesome murders and crime scene investigation techniques, and for taking time out of her busy schedule to visit Sweden and write a foreword to this book.

Friends and colleagues have read and commented on the manuscript in various stages, or contributed in other ways. We

The Rhubarb Connection and Other Revelations: The Everyday World of Metal Ions
By Lars Öhrström and Jacques Covès
© Lars Öhrström and Jacques Covès 2019
Published by the Royal Society of Chemistry, www.rsc.org

thankfully acknowledge: Michael O'Keeffe, Andrew Greene, Lotta Wiker, Roy Thomson, Anthony Covington, Scott Taylor, Krzysztof Rykaczewski, Helen Champness, Neil Champness, Kim Dunbar, Graham Lappin, Omar Yaghi, Deborah Kays, Mats Tiborn, Igor Huskic, Elizabeth Gardner, Marie Brigantini, Agne Gelotte, Craig Alan Huber, NHM London Staff, Robin Hansen, Stanislav Strekopytov, and Marielle Agbahoungbata.

Remaining errors are of course entirely of our own making.

Lars would especially like to thank:

The University of California – Berkeley for a Visiting Scholar opportunity in 2017, and Omar Yaghi and his research group for hosting me.

Friends, colleagues and students at Chalmers University of Technology for listening to my rambling stories, and their patience with my absence in mind and occasionally in body.

Chalmers Excellence Initiative Nano for a dissemination grant.

Jacques would especially like to thank:

The mobility program TOR granted by the French embassy and the French Institute of Sweden for a one-week visit scholarship to Gothenburg hosted by Lars and Nina, and Chalmers University of Technology. A trip to Stockholm gave me the opportunity to encounter most of the Nobel Prize laureates who appear in this book at the Nobel museum, and to admire the fantastic colours of the Vasa described in Chapter 10.

Frank Gabel, a biophysicist in my research group, who introduced me to pumpernickel bread.

Contents

The Rhubarb Connection and Other Revelations: The Everyday World of Metal Ions
By Lars Öhrström and Jacques Covès
© Lars Öhrström and Jacques Covès 2019
Published by the Royal Society of Chemistry, www.rsc.org

OVERTURE

Charlie Chaplin Goes for Gold but Gets Only Chromium

2013: The earth is stricken by a world-wide pandemic called the Scarlet Plague. In the San Francisco vicinity, a final hold-out in the chemistry department at a local university fails to protect the academics and their families, and the sole survivor, James Smith, wanders northwards. Back in the Bay Area some 60 years later Smith finds that his tales of 20th century technological marvels are simply taken as silly fairy tales in the hunter-gatherer society that has evolved among the very few survivors.

Probably because Smith, once a professor of English, hasn't much of a clue what is going on inside any of these wonders (for the record though, we know professors of Swedish literature who are also very able chemical engineers). Even less so of the chemistry hidden in a layer behind what even the normal engineer would see. A chemistry that, more often than not makes use of metal ions to help us do clever things, look smart, be frivolous and make a nuisance with chewing gum, keep vital, and keep criminals at bay. And by the way, it regularly saves lives too.

The Rhubarb Connection and Other Revelations: The Everyday World of Metal Ions
By Lars Öhrström and Jacques Covès
© Lars Öhrström and Jacques Covès 2019
Published by the Royal Society of Chemistry, www.rsc.org

But this book is not a clandestine chemistry textbook, the ambition is just to tell you how metal ions, in all their appearances and disguises, appear in the most unexpected circumstances. We will meet doctors shouting *Chelation therapy!* across clinics to get a life-saving treatment going, an emperor collecting non-precious metals in the wrong place, Oxford Dons harvesting hair from barbershops for their secret WWII project, con-men with their reoccurring gold-from-water swindles, and some ordinary people for whom various metal ions have played a life changing role.

Jack London's 1912 post-apocalyptic novel *The Scarlet Plague* (Figure 1) however, is devoid of real chemistry despite the main protagonist spending a lot of time in the Chemistry Department at the University of California in Berkeley. But of course, it could not have been written without it. Jack London, the most famous son of Oakland, San Francisco's less posh neighbour, did use a typewriter, but apparently first drafted his prose in longhand, presumably with a fountain pen. And a reliable formula for ink, used with different modifications

"POURING OUT OF THE CITY BY MILLIONS"

Figure 1   Fleeing the Scarlet Plague in 2013, illustration by Gordon Grant for Jack London's 1915 novel. Reproduced under the Project Gutenberg License.

ever since antiquity, involves both a chemical derived from the eponymous trees of London's home town and certain metal ions. Together they form pitch-black compounds giving a permanent colour on parchment or paper that is still a legal requirement for certain documents.

Known as iron gall ink, it combines the faintly coloured iron ions (like those you find in an iron supplement pill) with the likewise pale gall compounds to give something completely new with unexpected properties. An everyday thing that works because of metal ions, well hidden from the eyes of the casual observer.

A famous line from Jack London's hand was, however, written in pencil. Forensic experts (we will meet a few as we move along) have confirmed that the statement "Jack London, miner author, Jan. 27, 1898" found on the centre of a wall in the rear of a log cabin, was indeed written by the 22 year old, yet-to-be-famous author. Pencils use graphite, pure carbon, to trace a grey-blackish line, and as this book is about metal ions and their compounds, clearly, graphite does not qualify. However, the location of the log cabin, Henderson Creek in the Yukon Territory in Canada, does, because this is not far from Klondike where young London had joined the 1896 gold rush.[†]

Jack never made it big in the Yukon, but eventually became the best known and highest grossing author in the US. Whether London's *The Cry of the Wild* and other writings inspired Charlie Chaplin's 1925 *The Gold Rush* is unclear, although the two did meet before London's premature death in 1916. Gold features in the movie, but also a famous scene where The Tramp, Chaplin's legendary character, starving in a log cabin on the Yukon, cooks and eats his own shoes.

Leather is made up of proteins, basically, so this does make sense, but a major invention in tanning technology some 20–30 years prior to the Klondike Gold Rush means that by then, most leather would have contained a fairly-high amount of chromium as well. If you think that sounds unappetizing, you should consider the process which it replaced. Chomped brains and faeces

---

[†]Timber from the cabin was used to construct a new cabin in Jack London Square, Oakland, close to Heinold's First and Last Chance Saloon where Jack used to study to meet the entrance requirements for the University of California – Berkeley.

made the tanners both smelly social outcasts and exposed them to considerable biohazardous risks.

The Tramp's sturdy companion serves himself the tastier (one presumes) upper side leather, leaving the sole for The Tramp. This may or may not have been all right as far as the nutritional value goes. The sole might just have been of tougher leather, but in the late 19th century rubber technology was also moving into the shoe-and-boots business, although the invention of the rubber heel was still a few decades away (Figure 2). So, the sole could have been more of a chewing gum than a hearty meal. And rubber, even though it is a purely organic compound (or class of compounds), has plenty of reasons to appear in a book about metal ions.

The enzymes in the rubber tree need magnesium (Mg) or manganese (Mn) ions to glue together the smaller molecules that eventually will from rubber, and you need between 2000 and 15 000 of these to make one big rubber molecule. It turns out that the situation is similar when making synthetic rubber, but with different metals doing the job. We will talk about iron and

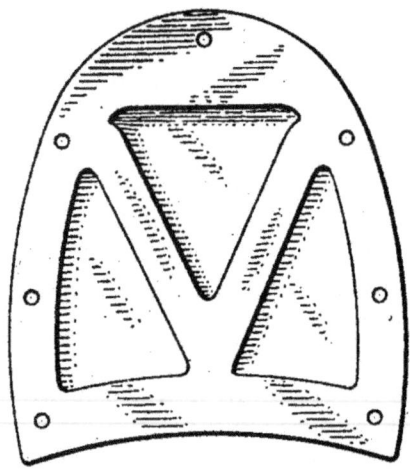

Figure 2   Illustration from the 1923 rubber heel patent by African American inventor Elijah McCoy after whom the expression "The Real McCoy" was minted. Rubber, although comprising only carbon (C), hydrogen (H) and an occasional sulfur (S), is nevertheless, as we shall see, a logical subject for a book about metal ions, their uses and their compounds. (E.McCoy Des.68 725 Rubber Heel, Filed Feb. 19, 1925, USD68725S).

lithium, as fortunately for this process you can avoid the much more expensive noble metals used in catalytic converters in cars, for example.

So, there is no gold involved in the rubber reaction, but loads to say about gold chemistry in other respects, some things more surprising than others. In another of Chaplin's classic movies, the 1936 *Modern Times*, The Tramp accidentally gets high on cocaine, and we will later see how CSI teams can use gold to detect this illegal drug.

Gold, however, we prefer to be used for jewellery. And while gold, silver and other metals, as they have only one kind of atom, might seem rather dull to a chemist once we have produced them (unless we are allowed to dissolve them and make new interesting compounds), the other component of jewellery, gemstones, is another matter entirely.

This gets us to the famous *Black Star of Queensland*, another type of star coming out of Los Angeles. In principle, a 733-carat piece of boring aluminium oxide of the corundum variety, a very common component of rock and gravel, and what protects most aluminium constructions from corrosion by forming a tight and thin resistant layer on top of the metal. Or rather, it would have been boring had the aluminium 3+ ions not in some places been replaced by tiny amounts of other positive ions of iron (Fe), titanium (Ti), chromium (Cr), copper (Cu), or magnesium (Mg).

Star sapphires come in a variety of colours depending on the trace metals in them, but all have in common an inclusion giving rise to a star-shaped reflection inside the stone. This occurs because of a kind of accidental mineral grafting. Just as you can graft closely related trees and plants onto each other, so you can make crystals that have common atoms and common structural features grow into one another.

In this case, as the sapphire crystal grows, two oxide ions attached to aluminium on the inside of the new crystal will grab a titanium 4+ ion instead of an aluminium 3+ ion on the outside. If enough titanium is present a new structure called rutile will start to grow.

This is a very different pattern from the sapphire structure, but can be grafted into the sapphire because in some directions in the crystal, the oxygen–oxygen distances are very close to the rutile oxygen–oxygen distances. But in other directions, the distance discrepancy is bigger, which means the rutile crystals will grow in a few selected directions only, and just in tiny needles, as there would otherwise be too much strain in the crystal and it would burst. These tiny needles reflect the light and also the internal symmetry of the crystal, giving rise to the star-shaped optical phenomenon shown in Figure 3.

The Black Star of Queensland itself acquired a rather troubled history once it left its home in the workshop of the famous Kazanjian brothers in Los Angeles. It started well, however, the stone "debuting" in New York in 1948 in the hands of Hollywood actress Linda Darnell, accompanied by flashing photographers and impressed reporters. This was the largest star sapphire ever, and it was subsequently seen around the neck of Cher in the 1970's as the Kazanjians had lent it to The Sonny & Cher Comedy Hour show on CBS. Drama followed when gemstone cutter Harry

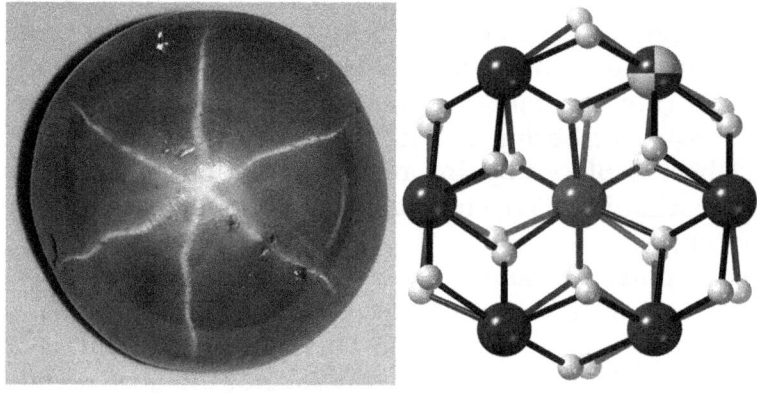

Figure 3   Left: Just like this blue variety (specimen is BM.89493 14 × 13 × 9 mm, photo credit: The Trustees of the Natural History Museum, London, reproduced under the terms of the CC BY 4.0 license http://creativecommons.org/licenses/by/4.0/) The Black Star of Queensland is a star sapphire, famously cut by Harry Kazanjian in Los Angeles in 1948. In principle, a boring piece of aluminium oxide of the corundum variety, the value created by trace amounts of other metal ions. Right: The corundum structure, black aluminium and white oxygen. The "beach ball" atom represents an aluminium ion that has been exchanged for another metal ion to give a sapphire (or ruby if red and containing chromium). The star shaped symmetry of the atom arrangement is mirrored in the reflection produced by the titanium oxide inclusion impurities in the Black Star of Queensland.

Kazanjian's nephew Michael had to spring up from the audience to rescue the stone when Cher started dancing.

Cher appears to be the only person to have ever worn the Black Star of Queensland in public, but the stone was later displayed at both the Royal Ontario Museum in Canada and the Smithsonian Museum in Washington.

It was in Washington that the trouble began, a visitor fell in love with the stone, and the Kazanjians sold it after more than 50 years in the family. The love story turned sour, however, when ownership of the stone was disputed between the buyer and his girlfriend, who apparently was the one that had dished out all the money.[1] The argument ended some years later with a vicious custody fight in the L.A. County Superior Court, and a quiet out-of-court settlement of an undisclosed sum. While its whereabouts for the moment seems unclear, the Black Star of Queensland was reported to have been up for sale in 2017, with an estimated value of $88 million. All for some trace impurities in aluminium oxide, one of the most common minerals on earth.[‡]

In the first chapter, we will continue with bullet hole detection kits and a cheating biker, but remember, we are not out to teach you chemistry, but to tell you in an entertaining way of the importance of chemistry in general and metal ions in particular.

If, however, you learn something in the process, it cannot be helped, and perhaps you will find yourself better prepared than the classic scholar James Smith, who bitterly complains at the end of Jack London's *The Scarlet Plague*: "If only one physicist or one chemist had survived! But it was not to be, and we have forgotten everything."

REFERENCE

1. V. Kim, For some, a sapphire has not been their best friend - The 733-carat Black Star of Queensland is at the center of an L.A. legal squabble that centers on allegations of deception, unkept promises and a lover's betrayal, Los Angeles Times, 2010.

[‡] The Kazanjian Red diamond has an even more adventurous story, but diamonds are made of carbon atoms only and therefore, unfortunately, outside the scope of this book.

Fantastic Metal Ions and How to Catch Them

In the Periodic Table of Elements there are 92 naturally occurring building blocks of our world and most of them are metals. Some give us power, other give us wealth, and a fair number just keeps us and all other living beings on this planet alive and well. This book is about how to catch them, put them in the right places, make them do good things and sometimes to remove them from places where they might do harm.

1.1 THE TARNISHED PLATINUM CYCLIST

Before starting in earnest we need to sort out some basic stuff, and we do this together with a rather questionable character, a young American who in July 1996, with the breath-taking speed of a racer cyclist, was on his way to Aix-les-Bains in south-east France. For the last few years, he had been pedalling himself into the world cycling elite, and now he could perhaps make his breakthrough in this, possibly the world's toughest athletic competition, the *Tour de France*. However, he did not make it to Aix, and a few months later it did not look like he would ever make it anywhere else on two wheels either. Lance Armstrong received

The Rhubarb Connection and Other Revelations: The Everyday World of Metal Ions
By Lars Öhrström and Jacques Covès
© Lars Öhrström and Jacques Covès 2019
Published by the Royal Society of Chemistry, www.rsc.org

frightening information from his doctors: he had testicular cancer and it had already spread to his lungs, abdomen and brain.
He was given a less than 50% chance of survival.

Followers of *Le Tour* and, indeed, sports in general, all know
how it ended. Armstrong recovered and went on to win a
record-breaking seven consecutive Tour de France competitions
from 1999–2005. Only to be disgraced and have them taken
away years later when the long controversy of his alleged use of
banned performance enhancing compounds finally reached the
end of the road.

Why are we telling you this you wonder? Because what saved
his life from the cancer was not only his tough physique and
possibly some luck, but a very simple compound that is an easy
first example of the type of chemistry that we will discuss in this
book, the block-buster drug cisplatin.[1] (To be clear, cisplatin is
not a performance enhancing drug.)

Discovered by serendipity and then developed by much hard
work by Barnett Rosenberg, Loretta Vancamp and co-workers at Michigan State University some 30 years earlier, it has
a fascinating history of its own. Right now, however, we are
concerned only with its simple components, two molecules
of ammonia, two chloride ions and one platinum atom in
the form of the ion with charge plus two, and how they stick
together to form this amazing drug molecule, formally named
cis-diamminodichloroplatinum(II).

That is a bit of a mouthful, but start by noting that ammonia
has the formula NH_3 and is the same molecule that we find in
some cleaning agents, that the chloride ion has charge minus one
and is identical to the chloride ion in table salt, NaCl, and that
platinum atoms are also found in catalytic converters for cars,
but this one has two electrons removed to give the 2+ ion. Figure
1.1 represents how we like to picture the cisplatin molecule.

Now, in your mind remove the bonds between Pt (platinum)
and N (nitrogen), and between Pt and Cl (chlorine), and you will
find five detached units, one metal ion and four other units that
we call *ligands*[†] because they can bind to a metal ion. We will talk
a lot about these ligands, so take note of this term.

[†]From the latin *ligare*, to bind. In biochemistry this term is used for a small molecule that
binds to a protein whether or not the binding site contains a metal ion.

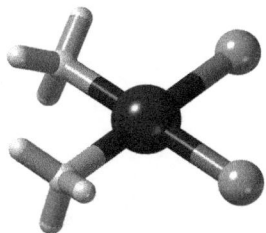

Figure 1.1   The cisplatin anti-cancer drug has improved testicular cancer survival so much that the diagnosis is no longer a death sentence. It consists of two ammonia molecules (NH_3 on the left side), a platinum 2+ ion in the middle (big and dark grey) and two chloride ions (Cl^-) to the right. This picture tries to emphasise the 3D-form of the molecule (note how the nitrogen and chloride atoms sit in the corners of a square) while still showing the bonds as spokes. We call this form "square planar" when we discuss how the ligands bind to platinum.

But do not fear, we will not pepper the text with unnecessary jargon. We do need to name things to be able to talk about them though, imagine the sport pages without naming the individual footballers? Full of "this fellow", "that fellow" and "the other fellow". That would really be confusing! But this is a trap popular science occasionally falls into, sometimes to the extent of making the text unintelligible to layperson and specialist alike.

Back to cisplatin, which unfortunately will not cure all cancers, and sometimes a cancer may also develop resistance[‡] against this drug. Then the doctor might suggest an alternative platinum-based drug, for example oxaliplatin (trade name elotaxin) which you can see in Figure 1.2.

If you now remove the bonds between the metal and the ligands you will see that we are now left with two ligands instead of four, and one metal ion. This is because each ligand now binds to platinum 2+ with two atoms each, and Pt^{2+} still prefers having four bonds arranged in a flat "square planar" geometry.

Of course, there is a name for this as well, but we are afraid you will not find the following very logical. We call a ligand that binds to a metal with more than one atom a *chelating* ligand,

---

[‡]This is not like bacterial resistance to antibiotics. This is an individual cancer in one person developing resistance, it cannot spread to cancers in other persons, so that unlike bacterial resistance, the rate of cisplatin resistance is the same now as when the drug was introduced by Bristol-Myers Squibb in 1978.

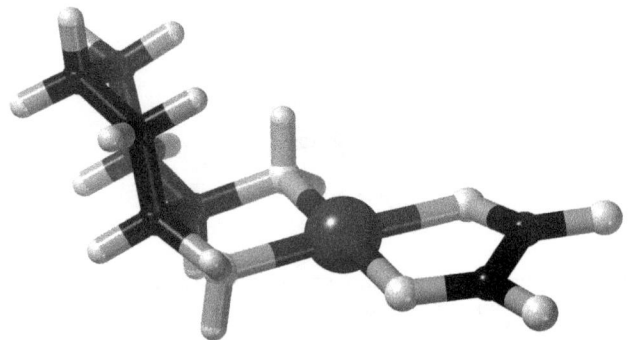

Figure 1.2   The anti-cancer drug oxaliplatin is used when cisplatin does not "bite". The large ligand with many carbons (black) and hydrogens (white) to the left is a man-made molecule, whereas the ligand to the right with the oxygens (white) is the oxalate ion, as found in many fruits and vegetables, notably in rhubarb.

and this comes from the Greek for claw, and what the coiners of this term thought about was the claws of a lobster as the ligand clearly pinches the metal and holds it fast. Fair enough, you might think, but then we call a ligand that can make a chelate a *multi-dentate ligand*, thus a ligand with many teeth, we count one for each atom that binds to the metal. And now it gets weird, as it is not clear if a lobster even has teeth, and if it does, they are certainly not sitting in the claws but seem somewhat surprisingly to be found in the stomach.

But here we are, and a claw with many teeth, incongruous as it may sound, is good if you want to hold on to a metal ion or catch it before it does harm or disappears before you can do something useful with it.

1.2   CHELATION CATCHES THE CRIMINAL

One standard use of this *chelate effect*, the way in which a multi-dentate ligand is much better at binding and "catching" a metal ion than two similarly looking but mono-dentate (and thus non-chelating) ligands, is in scene-of-crime forensic analysis. Lance Armstrong may have acted immorally, at least in the eyes of the sporting world, but he didn't wave guns around, and there were no bullets to look for.

However, at a crime scene where shooting is suspected, this is exactly what you are searching for, and if you find a bullet hole,

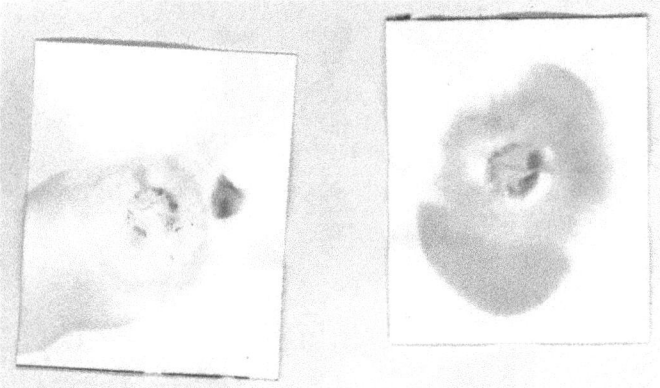

Figure 1.3   Scene of crime bullet hole detection. The paper has been dosed with a detection liquid and pressed against the suspected bullet hole. A rose colour indicates the presence of lead and black-green indicates that copper has left a trace on impact. In black-and-white we can only see the circular traces. Photo credit: Lars Öhrström, holding the gun and detecting the bullets: Inspector Marie Brigantini, Swedish Police West.

you will also stand a good chance of finding the actual bullet that can then subsequently be tied to a certain weapon. In some cases, it is not obvious which holes are bullet holes and which are not. Or worse, in some cases, it might be hard to know which are new bullet holes and which are old. That is why a small attaché case with a bullet-hole-test-kit is always carried to the scene of crime. The ligands in the single-use vials will both catch the lead or copper ions present in the bullet traces left in the hole and give a distinct colouring that can be detected on the spot (see Figures 1.3 and 1.4).

Having said that, nothing is ever as simple as implied in the high-heeled-CSI shows on the telly. Ambiguous results, false positives and failure for the reagents to be specific enough, or to react with a low enough concentration of the suspected criminal residue, are some problems facing both the investigating officers and the chemist developing the methods.

1.3   WHAT ABOUT THE METALS?

So, multi-dentate ligands can catch metal ions, but there are so many different metals, are there no differences between them? Of course there are. Copper for example, likes sulfur (S) a lot,

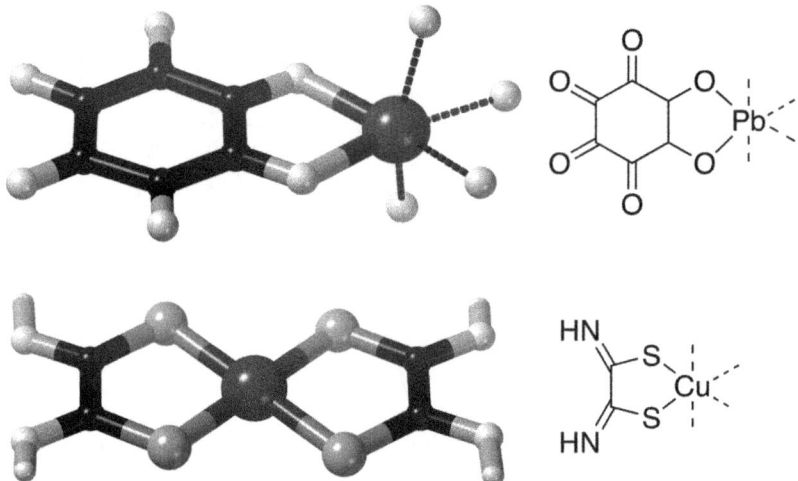

Figure 1.4 Bullet hole detection: The bi-dentate ligands form chelates with the metal ions Pb^{2+} (left, only one ligand $C_6O_6{}^{2-}$ shown, the remaining bonds are represented by dotted lines) or Cu^{2+}(centre, both dithiooxamide ligands $H_2C_2S_2{}^{2-}$ are shown, this complex is charged 2+) giving coloured compounds that can be detected on the spot by the naked eye, allowing the forensic examiner to go on to find the bullet. Right: The corresponding so-called Lewis-structures. These will occasionally be displayed for the benefit of those needing some more detailed information.

and it is not a coincidence that the Cu^{2+} ion is detected with a sulfur containing ligand in the bullet-hole-test. Calcium ions (Ca^{2+}) on the other hand, do not have much affinity, as we say, for ligands with sulfur, as their bonding atoms prefer oxygen, which is their natural surroundings in bones, teeth and marble for example. The metal ions also come in different sizes, the Li^+ ion, very important in the treatment of some forms of bipolar disorder, is very small, 73 picometres, whereas the Pb^{2+} ion, a rather nasty fellow, has almost twice this radius, 143 picometres, that is 0.000000000143 millimetres.

It will not be necessary to memorise the Periodic Table to follow the remainder of the book and for now, a quick look at Figure 1.5 will suffice.

There are so many metals, but still fewer than footballers in the Premier League. We will be mostly concerned with some of the "stars" such as gold (Au), but also lesser known metals, such as gadolinium (Gd) might find their way into the stories.

We do not want to be biased though, every naturally occurring element is important in some way. As a student once said "all elements are created equal and should be treated with respect" (this particular chemistry class had been preceded by a diversity workshop).

One thing to take home from Figure 1.5 however, is that elements appearing in the same column tend to have similar properties, that is kind of why we call it the periodic table. Therefore, lithium and sodium will both form ions with the charge +1, calcium and strontium (Sr) will be 2+, and silver (Ag) and gold are noble metals, just to take a few examples.

On the side of the non-metals, we have the atoms that make up our ligands, and here we will be more restrictive. We will almost exclusively stick to carbon (C), hydrogen (H), oxygen (O), nitrogen (N), sulfur (S) and phosphorous (P), with the occasional chloride (Cl) or bromide (Br) ion making a guest appearance.

Now, this may have been a bit dry and reminded you of your school days. If so, rest assured that the rest of the book will contain more action. I'm sure you realise that to appreciate a game of football, you need to know the basic rules even though you might not grasp the finer details of the offside concept, and the same is true for chemistry.

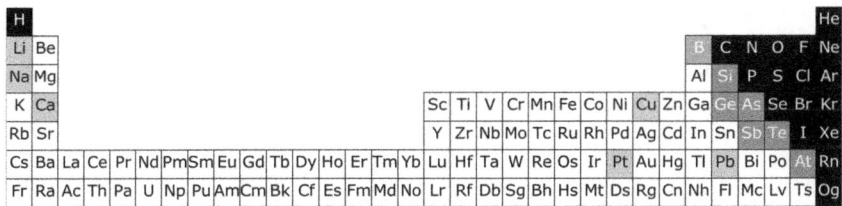

Figure 1.5 The Periodic Table of Elements.[§] On the black background are the non-metallic elements, the metals are on the white background, and on the grey background with white text are the elements that sometimes behave as metals and sometimes not. The six metals we have already encountered, lithium (Li), sodium (Na), calcium (Ca), copper (Cu), platinum (Pt) and lead (Pb) have black symbols on a grey background.

[§]The IUPAC (International Union of Pure and Applied Chemistry) released the table on November 28th, 2016 (https://iupac.org/what-we-do/periodic-table-of-elements/). Often the two rows La–Yb and Ac–No are cut out and placed below the main table, but here you can count the atomic numbers from 1 for hydrogen (H) to 118 for oganesson (Og) from the upper left corner to the bottom right corner.

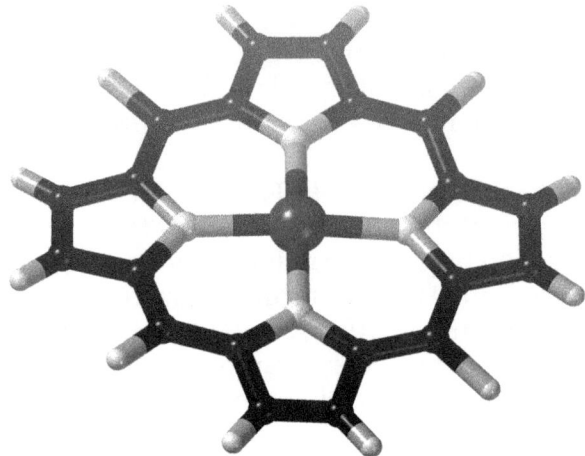

Figure 1.6   A model of the iron containing heme group in haemoglobin. Iron
is trapped in the centre by the four nitrogen atoms of a porphyrin
ring which is what we call the central part of the heme molecule.
In our body the porphyrin ring has a number of different additional groups dangling from it.

Although the next chapter is going to be completely inorganic,
save for some rhubarb, a lot of the stuff in this book relates to
biochemistry and living organisms, from fascinating sea-snails
to humans. Let us therefore finish by just showing a molecule
you have in your body and that we will discuss further later on,
the heme group in haemoglobin. This is an example of how efficiently some biological molecules trap metal ions in a perfect
cavity between four nitrogen atoms arranged in a square with the
metal at the centre.

The molecule works like a kind of a molecular lasso, but
instead of a cow, we have trapped iron in the heme-molecule that
is attached to our haemoglobin and without which we cannot
breathe (Figure 1.6).

However, the oxalate ion in Figure 1.2 and the bullet-hole detection ligands in Figure 1.4 look more like pincers than lassos, but
may be equally efficient, as we shall see in the next chapter.

REFERENCE

1. J. Wennerberg, *Läkemedel som förändrat världen: historier om
vetenskap, slump och envishet*, Apotekarsocieteten, Stockholm,
2012.

Rhubarb, an Emperor and the Butler's Pride

September 1870 had already started badly for Charles Louis Napoléon Bonaparte, and it was going to get worse. At Sedan, a small French town close to the Belgian border, the commander of the French army, Patrice de Mac-Mahon, was wounded and the sick and weak emperor, better known as Napoléon III and nephew of the great Corsican, had to surrender and give himself up as a prisoner to the invincible German duo, Helmuth von Moltke and Otto von Bismarck.

In the run-up to the battle of Sedan, the decisive battle of the Franco-Prussian war of 1870–1871, von Moltke had successfully engaged the French army by a classic pincer movement, attacking from two sides rather than in the centre. Urate ions, that have nothing to do with the element uranium, but are a product of the break-down of DNA, attack sodium ions, Na^+, and calcium ions, Ca^{2+}, in the same way. Two electron pairs on each of the negative oxygen and nitrogen atoms can bind, from two sides, to any metal ion that is in its way (Figure 2.1).

As long as these compounds float around dissolved in our bodily fluids they are not a problem, but when they become too

The Rhubarb Connection and Other Revelations: The Everyday World of Metal Ions
By Lars Öhrström and Jacques Covès
© Lars Öhrström and Jacques Covès 2019
Published by the Royal Society of Chemistry, www.rsc.org

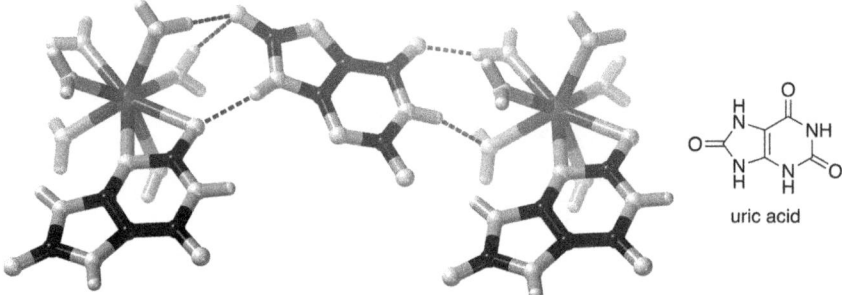

uric acid

Figure 2.1 Urate ions (fused hexa- and pentagons) binding to calcium ions, Ca^{2+}. They occur naturally in our body and are the anions of uric acid, the compound has a H+ ion attached to the negatively charged nitrogen where the calcium 2+ now binds. Each calcium also binds six water molecules (the V-shapes bonded to grey calcium) and the dotted line indicates the so-called hydrogen bonds, the same bond that holds H_2O molecules together in ice, snow and water.

concentrated and starts to add onto each other *via* a second pincher attack on the backside of the sodium ion, or the flat urate molecules start to stack one upon the other, we might have a problem. As now we will start to form particles, and just as a snowball, they will pick up more and more of the urate and chelate even more metal ions until the "stones" are so big that this becomes a very painful condition.

2.1 THE EMPEROR'S STONES

Such stones were something Napoléon III was all too familiar with, and during the battle of Sedan he was in terrible pain, owing to his stones (Figure 2.2). Was the war lost because of this? Probably not; it seems the war was lost as soon as it was declared because of the Prussian technical and organisational superiority.

However, the Emperor was not very keen to take up arms in the first place, with some of his advisors being the chief warmongers. How well he could resist and counteract them, while being beset by his bladder stones, is difficult to know.

It is easier to turn to the "stones" themselves, because these are susceptible to detailed methods of scientific interrogation

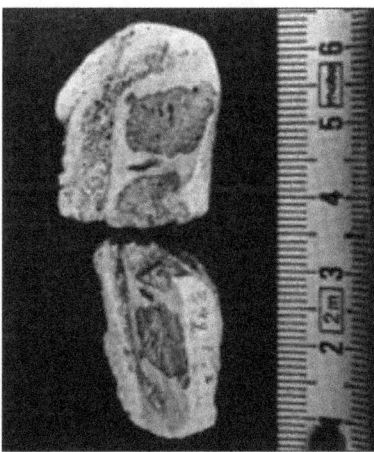

Figure 2.2   The "stones" of Napoléon III, removed after his autopsy in 1873. (Scale in centimetres.)

such as X-ray crystallography. This is the supreme method we use to find out the detailed arrangement of atoms and molecules in solid materials and other chemical compounds,[†] and these could be anything from metals to proteins or to somebody's kidney or bladder stones.

Surprisingly there are things we still do not know about the urate compounds, and detailed information is just emerging. Such as the structure of the first calcium compound reported by Janeth Presores, Jennifer Swift and co-workers in 2013, and the sodium urate structure found by analysing the stones coming from a 63 year old male.[1,2] In the latter case, Dominique Bazin and his team noted that this solved a 150 year-old problem, but that they needed the most up-to date technology to do so, namely a large-scale facility known as a synchrotron. Somewhat confusingly these solid objects are known as "calculus" in medicine, not to be confused with the branch of mathematics that some people unfortunately seem to find equally painful, and that we might just need to avoid this painful condition.

[†]Also known as X-ray diffraction and associated with many Nobel prizes. Subject of the UNESCO sponsored International Year of Crystallography in 2014. See https://www. iycr2014.org.

2.2   THE RHUBARB CONNECTION

Urate ions are not the only organic component of these painful, and sometimes dangerous, stones, (Napoléon III died in the
aftermath of an operation trying to remove them) and a more
well studied component is that formed by Ca^{2+} and $C_2O_4^{2-}$ or
$^-OOC-COO^-$. This is the oxalate ion that we briefly saw as part of
the anti-cancer drug oxaliplatin in the preceding chapter, and the
compound, conveniently enough, is known simply as calcium
oxalate.

The oxalate anions are what you get, together with H^+ ions,
when you dissolve crystals of oxalic acid in water. Oxalic acid is
a small molecule with the formula $C_2O_4H_2$ but is perhaps better
represented as HOOC–COOH; two carboxylic acid groups joined
together to make the simplest organic di-acid possible. You may
know this chemical as the compound that gives the taste of rhubarb that special acidic tang, but it is also found in spinach and
a number of other veggies that most of us happily eat. It is also
part of our normal metabolism and, just as urate, occurs completely naturally in our bodies.

Oxalic acid is named after a plant, the Common Wood Sorrel,
a small herb whose leaves have a distinct acidic taste, which can
be revitalising if you get lost in the woods, and has the Latin
name *Oxalis acetosella*. The most distinctive feature of this small
molecule is its shape. It looks a bit like two Y's pieced together
by overlapping their bases. Carbon atoms are at the two inter-
sections and oxygen atoms at the four ends, with the protons
(or H^+ ions) hanging one-by-one on either one of these oxygens.
Moreover, it is completely flat and when it has lost its protons
to become the oxalate anion, it has minus charges located on
two of its oxygen atoms, convenient for a pincher manoeuvre
(Figure 2.3).

Most technical applications of oxalic acid hang on its flatness,
negative charge, and ability to "pinch" metal ions in a chelate as
a bidentate ligand. The positive calcium ions love the more negative oxalate oxygen atoms, and to such an extent that they do not
care if the oxalate oxygen atoms carry a negative charge or not.
So not only will there be a pincer movement forward towards one
calcium ion, but also backwards, attaching to another one in that
direction. As each calcium ion is rather big, four oxalate ions will

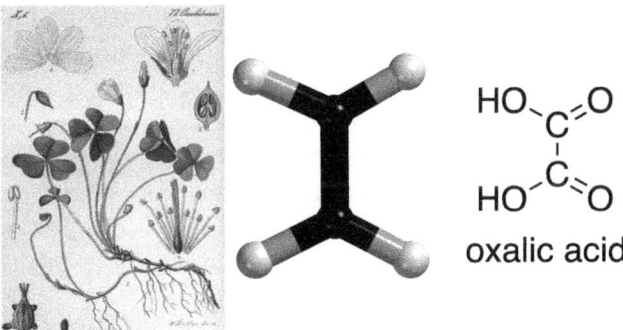

Figure 2.3   Left: *Oxalis acetosella*, the Common Wood Sorrel by Otto Wilhelm
Thomé in Flora von Deutschland Österreich und der Schweiz,
painted 15 years after the faithful battle of Sedan. Middle: ball-
and-stick model of the oxalate ion, this remains once oxalic acid
has given away two H^+ ions. Right: The Lewis structure of the par-
ent oxalic acid.

assemble around each metal ion, each attaching to two other calcium ions in the other direction. In this way calcium oxalate will
grow into an infinite network that will be completely insoluble
in water. This is the most common cause of kidney or bladder
stones, and Napoléon III probably also had some calcium oxalate
intergrown with his urate crystals.

2.3   EXCUSE ME, THERE IS AN EQUATION ON THIS PAGE!

Yes, this is where the calculus comes in so that we can avoid the
calculus! At this point we need to know under what conditions
the calcium oxalate, or any of the other "stone" forming minerals,
start to grow crystals instead of just happily swimming around in
your body fluids. Is it when we have exactly the same amounts
of each of the parts? Or when one of them exceeds the other by
some fixed amount? It turns out that as long as the product of
the Ca^{2+} concentration and the oxalate concentration does not
exceed a certain, rather low, value, both ions will stay in solution.
We express this as:

$$[Ca^{2+}] \cdot [\text{oxalate}] = K_{sp}$$

Inside the [...] we have the value of the concentrations, and
K_{sp}, the solubility product constant, is a very small number, to be

more precise 2.32×10^{-9}, in other figures 0.00000000232. You can compare this to having permission to build a house with a specific maximum square meter surface. You can make it any width or length, as long as the product of the two does not exceed the allowed square meter surface. If they do, then you will be in trouble, and the same goes for prolonged periods of exceeding what we call the solubility product limit of these stone-forming substances, because then stones will start forming in your bladder or your kidney.

But do not be alarmed, apart from in some very special medical conditions, a normal, or even high, intake of calcium ions and oxalic acid from natural sources does not seem to increase the risk of kidney stones.

Oxalic acid can also be produced in the body if the antifreeze compound ethylene glycol ($HOCH_2CH_2OH$) is consumed by mistake, or by a deliberate poisoning attempt. By a chain of events triggered by the enzyme alcohol dehydrogenase doing its work on ethylene glycol instead of ethanol, four unpleasant compounds will be produced, ending with oxalic acid. The detection of calcium oxalate crystals in the urine may therefore also indicate attempted murder by the easily accessible chemical ethylene glycol.[‡]

2.4   THE BUTLER'S PRIDE

Mr Stevens reminiscences with pride "...that the state of the silver had made a small, but significant contribution towards easing of relations between Lord Halifax and Herr Ribbentrop that evening." Which, he notes a bit further on, made "...a contribution to the course of history." But the world where gleaming silverware took on such an importance was disappearing by the time he is telling his tale, evidenced by the closed site of Giffen & Co, makers of the ultimate silver polish, that Stevens passes on his way to see his former colleague Miss Kenton.

Silverware and its states occur now and then in Kazuo Ishiguro's 1989 novel *The Remains of the Day*, and for anyone that has ever visited a British "stately home", or a UK regimental museum, it

---

[‡]More details on this and on an attempted Christmas Day murder can be read in Raychelle Burks short piece *It's the most poisonous time of the year* https://www.chemistryworld. com/opinion/have-yourself-a-deadly-little-christmas/3008320.article.

is clear that we are not only talking about objects would that normally find use on a high-end dinner table, but also of an assortment of extravagant and sometimes outlandish ornamental pieces.[§] In Ishiguro's famous novel, adapted for the cinema with Anthony Hopkins and Emma Thompson as Mr Stevens and Miss Kenton, the diminishing role of silver and its polishing is also symbolic. Darlington Hall has a new master, and Stevens new American employer does not care much for silver or its potential in amateur diplomacy. The old world is disappearing, and no amount of polishing can recreate the lost opportunities of 20 years ago. The shining surface may only faintly reflect back to you what might have been.

Having said this, we do occasionally polish silverware even today, provided we are bourgeois enough to both have them and care, and we still use products close to Giffen & Co's, despite 90 years of chemical development and numerous quick-fix methods described on the internet. We now cite directly from the latest edition of *Ullmann's Encyclopedia of Industrial Chemistry*:
"The most effective method of treatment, however, is still the laborious process of cleaning with a paste, which removes not only the surface layer of silver sulphide but also small surface imperfections, giving a smooth polish and restoring the original luster and warm color of the silver."

In contrast to the fictional firm of Giffen & Co, the makers of Silvio polish still seem to be going strong, as the logo and trademark is little changed since the 1920's.[¶] It is, however, notoriously difficult, despite consumer rights laws, to know the exact composition of these concoctions.

Our nose, which in some respect is quite a sensitive analytical instrument, gives use some hints though. The characteristic smell of silver polish comes from sulfur compounds, such as octadecyl-3-mercaptopropionate and other long chained hydro-carbon compounds with a sulfur atom close to one end. (Mercapto

---

[§] A famous silver object from another brilliant author's career is a Cow Creamer, the stealing of which is reluctantly pursued by Bertie Wooster and his valet Jeeves in another stately home, Totleigh Towers, on orders from his aunt Dahlia in *The Code of the Woosters*, by P.G. Wodehouse (Herbert Jenkins, Doubleday, Doran, London, 1938).

[¶] The makers, Reckitt and Sons are now, after a series of mergers and splits, part of Reckitt Benckiser but still (in 2017) retain substantial activities in the town of origin, Hull in Yorkshire.

has nothing to do with the element mercury but signals the presence of a sulfur atom.)

The logic of this is that the tight-fitting, but very thin, and black, surface layer of silver sulfide (Ag_2S) has many Ag–S bonds. These are very strong, and thus silver sulfide is very insoluble, much as the dreaded kidney stones. The reason the sulfide coating cannot be washed away with water, is that the Ag–O bonds that then would form are much weaker than the Ag–S bonds. Instead, we need to replace these with other Ag–S bonds if we are to remove the silver ions and with them the ugly looking surface layer, and long-tailed mercapto compounds[‖] are just the ticket.

Why long tailed ones you might ask? Because the more $-CH_2-$ units we add to a hydrocarbon chain, the less the probability for it to leave the ground and find its way into your nose. Or, to put it more scientifically, they are less volatile and have higher boiling points. This is a good thing indeed when we are dealing with sulfur compounds as these are notoriously smelly.

The sulfur atom has six electrons in its outermost shell, the same as oxygen (that is why it is placed just under oxygen in the Periodic Table) and uses two of these to make shared two-electron bonds with carbons, making four electrons available to catch silver ions. However, this is not all, both silver and sulfur atoms are classified as being "soft" because they have many electrons and are big. This means their outermost electrons are not so tightly held by the positive nucleus[**] and these electrons can readily rearrange themselves in interactions with electrons from another soft atom to form a stronger chemical bond. This is not the case with a small and hard atom like oxygen, where the electrons are held very tightly and have little freedom to move around. This is why the Ag–O bond is less strong, Ag is soft and O is hard.

Still, we associate silver polishing with some kind of odour sensation, and that is because our nose is very good at detecting

---

[‖]Like octadecyl-3-mercaptopropionate.

[**]You might recall from school that the force attracting a charged particle to another charged particle is proportional to the charges multiplied by each other divided by the square of the distance, $F = kq_1q_2/r^2$ so that the further q_1 comes from q_2 (r gets bigger) the weaker is the force, F, holding them together.

even very low concentrations of these compounds. On a molecular level, this is not yet completely mapped out, but we know that trace amounts of sulfur compounds in wine, beer, cheese, onions, grapefruit, and coffee, just to mention a few tasty foodstuffs, can be detected by us, and are very important for our overall appreciation of these things.

To make things more complicated, it is not only the ability of the sulfur-containing molecule to move from a liquid or solid state into a gas, but also its size and shape that matters. Specific proteins, large molecules made up from amino acids and additional molecular components, capture these compounds one molecule at the time, and the slight shape change this triggers in the protein sends a signal to our brain.

Now we seem to be digressing, but at least one of these proteins we use, the human odorant receptor OR2T11, has a metal ion in its binding pocket, and without that metal ion, we cannot detect the smell, or indeed, sublime taste, of these compounds. And, just as for the silver polish, it is the match between the soft sulfur and soft metal that does the trick, in this case not silver, but copper ions (one step up from silver in the periodic table, so we should expect the properties to be similar). With some chemical tricks one can even show that if copper is changed for silver,[††] the protein still works, whereas putting, for example, manganese (Mn) there instead, leaves the protein unable to detect anything.

The same idea is used in other metal polishes, and also to remove unwanted iron compounds, especially rust, from kitchenware and even marble. For iron, it is again the ubiquitous oxalate ion that is used, and this useful component is often also added to other metal polishes as well. We call them scavengers, and as such oxalic acid or oxalates are components in many technical cleaning formulations and these solutions all work on the assumption that the metal-oxalate compounds formed are water-soluble and thus can easily be washed away. This is also why your steel saucepan might look a bit shinier after preparing rhubarb compote in it!

[††]Note, however, that silver has no known biological role in our bodies, and the research was conducted on chemically isolated proteins in test-tubes.

2.5 A SIBERIAN GEOMETRY LESSON

The soluble oxalates generated by such rust removal will contain an iron 3+, or occasionally iron 2+, ion surrounded by three oxalate ions giving large ions with a net negative charge of 3− or 4−. By making neutral compounds of these with charge balancing potassium or sodium ions (we call these salts) one gets the type of compounds that can be washed away when rust is removed, but more importantly, chemists consider these as beautiful compounds owing to their peculiar propeller like symmetry. In 2016, a mineral from Siberia was found to contain these iron oxalates with the charge partially balanced by magnesium ions surrounded by six water molecules, $[Mg(H_2O)_6]^{2+}$,[‡‡] and in Figure 2.4 you can see both the beautiful green mineral and part of its equally beautiful crystal structure.

It should be noted that all metal ions are bonded by six oxygen atoms. If we make lines between the binding oxygens we will form an octahedron, one of the five Platonic solids, and we simply call these types of metal compounds octahedral.[§§] This

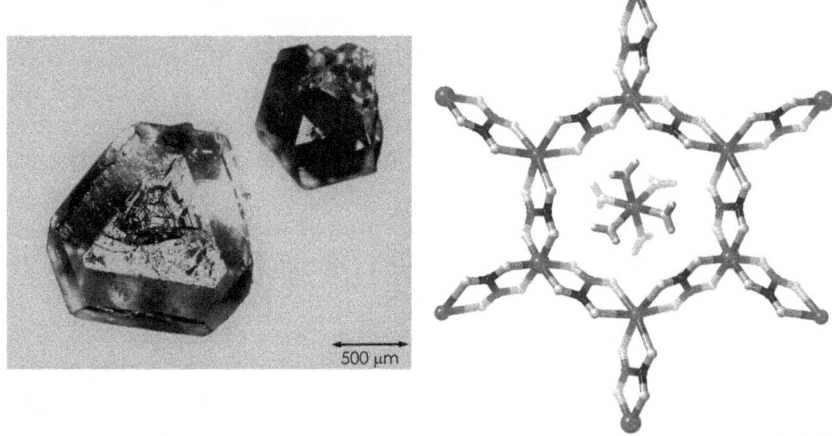

Figure 2.4   Crystals of synthetic stepanovite (Photo Igor Huskic, reproduced with permission) and part of its crystal structure. The naturally occurring stepanovite is found in Siberia.

[‡‡]When we write formulas, we put the metal ion and its ligands inside square brackets.

[§§]A Platonic solid has all faces and vertices equal. It may seem that chemists are obsessed with symmetry and beauty, both for aesthetic and practical reasons, and much has been written about this, notably by chemistry Nobel laureate and poet Roald Hoffmann, see for example Tami Spector, *Of atoms and aesthetics*, Chemistry World, July, 2014. Approaching from a different direction, Dutch graphic artist M.C. Escher independently derived important mathematical symmetry rules used by chemists in crystallography (used to obtain the structure in Figure 2.4 and many others in this book) and put them to work in his own art.

is by far the most common way for a metal ion to interact with neighbouring molecules, be they water or something else, and we will see this coordination geometry as we call it, frequently in the following chapters.

The pincher movement used by the oxalate ions in this compound, so fatal to Napoléon III, his army and his calcium ions, will be put to good use in the next chapter.

FURTHER READING

The rhubarb plant itself has a fascinating history, see for example: I. Nilsson, *Rabarber*, Bokförlaget Arena, Lund, Sweden, 2016.

REFERENCES

1. D. Bazin, M. Daudon, E. Elkaim, A. Le Bail and Ĺ. Smrčok, *C. R. Chim.*, 2016, **19**, 1535–1541.
2. J. B. Presores, K. E. Cromer, C. Capacci-Daniel and J. A. Swift, *Cryst. Growth Des.*, 2013, **13**, 5162–5164.

CHAPTER 3

Mining, Magnets and Making New Elements

A gentle hiss can still be heard from the lamps in Agnes Toward's flat in the tenement on Garnethill. When she moved into this typical Glaswegian flat in 1911, the gaslights had still not been replaced by electric lights, and the attraction of the flat, now a museum run by the National Trust for Scotland, is her collection of everyday objects, meticulous cared for until her death in 1965. A time-capsule of early 20th century life (Figure 3.1).

3.1 MORE THAN MEETS THE EYE

Today, the brief era of gas light is all but forgotten, and Edison and the electric lightbulb is what clings to our memory, to the extent that getting a bright idea is often illustrated by a lightbulb appearing over someone's head. Now, if you look at it closely, the Edison story isn't so watertight, and we would argue that the real game changer was the gaslight. Electric light was just another, more convenient but essentially similar, solution to the same problem. A remotely generated energy carrier (chemical or electrical) was transmitted through pipes to a large number of end users where the energy was transformed first to heat and then,

The Rhubarb Connection and Other Revelations: The Everyday World of Metal Ions
By Lars Öhrström and Jacques Covès
© Lars Öhrström and Jacques Covès 2019
Published by the Royal Society of Chemistry, www.rsc.org

Figure 3.1   Agnes Toward's gaslights in Glasgow. Photo credit National Trust for Scotland, reproduced with permission.

in various proportions, to light by a very hot chemical substance that starts to glow but is not consumed in the process.

In Edison's (we call it this for simplicity) incandescent light bulb the substance was carbon in various forms, notably improved by the Afro American inventor Lewis Latimer. To prevent the carbon from burning and forming carbon dioxide and water, the bulb was filled with an inert gas or evacuated so that no oxygen was present.

But what about the gas light filament? It is not the flame of the gas that gives the light, it's another substance heated up, just as in the light bulb. In gas lights we call the filament mantles and nowadays we commonly find them in shops that sell outdoor equipment. This is because they are used in camping lights of various kinds, both gas and kerosene fuelled. But why on earth are we told in the shop that they are not radioactive?

Sometimes these days, a solid sales argument seems to be what is not in a particular product, irrespectively if it was ever there, or if it is reasonable to think that it should have been. However, for the gas mantles it is true that they used to emit fair doses of ionising radiation, the sign of radioactive compounds, and you can still find ones that do.

The reason? The standard way of making them used to be to impregnate a meshed cotton fabric with a thorium nitrate solution, and thorium (Th) is one of the most radioactive of the naturally occurring elements of the periodic table. Together with uranium, obviously, but there are also other elements that naturally emit ionising radiation. For example, the rare-earth praseodymium, Pr, but also the, for humans, essential element potassium, K, making nutritious potassium rich bananas slightly radioactive.[†]

As for the gas mantles, in the first use, the flame would burn off the organic material and the nitrate, leaving a thin thorium oxide mesh behind. When heated by the flame, the thorium oxide would glow with an intense pleasant white light, much as a LED light today converts electricity to light without getting hot. In comparison old light bulbs would turn extremely hot to the touch, converting the electrical energy mostly into heat and only a fraction into light.

These mantles are also known as Welsbach mantles after their Austrian inventor Auer von Welsbach, and used to be the standard material for gas mantles in, for example, streetlights. This was the major use of thorium until electric light became common. Even for everyday use they seem to be quite safe and will not cause a significant increase in the dose of ionising radiation that we all receive each day from natural sources, such as from potassium ions in bananas. The problem lies more with their manufacture, and the risk for the workers in these plants. It is therefore a good idea to stick to the non-radioactive mantles for your camping equipment in which thorium has been replaced by yttrium.

Now, what has all this to do with ligands binding to metals? Nothing to do with the oxides in the gas mantles, but we find the ligands instead in the process of extracting the metals from their ores.

3.2   SEPARATING THE RARE FROM THE RADIOACTIVE

It has been said that a mine is a hole in the ground with a chemical factory on top, and this is true even for gold that is mined as a relatively pure metal. All other metals need additional electrons to turn their positive ions into the metallic state, they need to be *reduced* as we say. For our most abundant construction metal, the alloy we call steel, which is mostly iron, these electrons come

---

[†]The banana equivalent dose (BED) is an informal measurement of ionizing radiation exposure.

from carbon in coal. This carbon in turn becomes *oxidized* to carbon dioxide, meaning that for every two metallic iron atoms that are formed from the ore, a steel mill will emit three CO_2 molecules into the atmosphere.[‡]

In addition, the mined metal ore will contain a larger or smaller number of undesirable elements from the periodic table, either as marriage partners in chemical compounds, copper for example is often attached to sulfur, or as impurities. These need to be eliminated in an efficient way and perhaps, as in the case of sulfur, refined and processed to other products. It should come as no surprise then that nowadays, a sulfuric acid factory is often added to the copper works.

So, at the end of the day, when physical methods like sifting, using magnetism, or density differences, have been exhausted and produced a crude raw material, it is always down to the chemists to deliver the final product.

For yttrium, and other elements known as the rare earths, or REEs, this very often involves a step in which a crude solution of positive metal ions of the element in question is reacted with oxalic acid. The oxalate ion grabs one metal ion with its front claws and another with is back claws. Then two, three or even more oxalate ions can fit around theses larger metal ions, each taking a "bite" without interfering with the others. Just like a gigantic Lego set where pieces are added in all directions, this gives huge structures, we could say that the metal ion and the oxalates polymerise, and these structures will be so large that they are insoluble and can conveniently be collected in the bottom of the reaction vessel leaving the unwanted elements behind in the solution. So they form insoluble compounds, much like gallstones.

The fact that many rare earth elements are not very rare does not mean that such mines are found in every corner of the world.[§] Because even if the refinery process is good and gives pure

---

[‡]Unless we can make use of hydrogen gas as a reductant instead, but we'll still need some carbon atoms to make steel out of iron.

[§]According to IUPAC the REEs are: lanthanum (La), cerium (Ce), praseodymium (Pr), neodymium (Nd), promethium (Pm), samarium (Sm), europium (Eu), gadolinium (Gd), terbium (Tb), dysprosium (Dy), holmium (Ho), erbium (Er), thulium (Tm), ytterbium (Yb) and lutetium (Lu), often with the addition of the group 3 elements scandium (Sc) and yttrium (Y). The rarest of these, except the radioactive promethium, are lutetium and thulium, and even these are almost 200 times more abundant than gold.

yttrium, neodymium, or gadolinium (this one we will encounter later in Chapter 7), one is left with the problem of the waste. It is this kind of waste that some people would rather see in someone else's back yard, as the rare earth elements are often associated with the radioactive rouges of the periodic table, thorium and uranium (U).

The reason is that these metal ions tend to have similar properties, charges and sizes. Unfortunately, they do not occur in high enough concentrations for it to be economically viable (for the moment one might add) to extract them out as a by-product of the REE production, but they still occur in concentrations high enough to be environmentally problematic in the waste stream.

For this reason, it has been argued that wind power plants, if their electric generators use permanent neodymium magnets, produce more radioactive waste per electric energy unit generated than a nuclear power plant. However, the problems generated by the two types of waste give rise to quite different technical problems and risks, so we are really comparing apples to bananas, and this is something that requires a more detailed analysis.

The REE production is so problematic that the only mine outside the People's Republic of China, Mountain Pass in California, was closed down for exactly these reasons in 2002, and gave China a virtual monopoly on rare earths.[1] This is a fact that the West and Japan became painfully aware of during a brief Chinese production decrease, leading to the REE crisis of 2010–2011.

3.3 EXQUISITE PHOTOGRAPHS

Even though digital photography is now rapidly replacing traditional chemical-based photography, or rather the single use chemicals are replaced by multiple-use registration-and-display solid state materials chemistry, in photography as an art form there is still place for traditional chemistry.[2]

One of the more exclusive printing methods is based on various photosensitive oxalate compounds with the iron 3+ ion mixed with simple platinum salts, usually a negatively charged $[PtCl_4]^{2-}$ anion with the four chlorides arranged in a square planar fashion, in combination with two sodium or potassium +1 cations.

Also, a combination of palladium (Pd) and platinum is possible as palladium has a very similar chemistry to platinum, its sister element that is situated one step below in the periodic table.

As you might have noticed, the oxalate ion can be written as $^-O_2C-CO_2^-$ where the "–" between the two carbon atoms represents a shared 2-electron chemical bond. What happens if we remove these two electrons from the oxalate? $^-O_2C-CO_2^-$ becomes $O_2C + CO_2 + 2$ electrons, and we can frivolously regard the oxalate ion as two carbon dioxide molecules in waiting, with two electrons to give away. In iron(III) oxalates, on the other hand, we have iron ions charged plus three, in principle ready and willing to take up one electron to form iron 2+ ions.

This does not quite add up, and an oxalate metamorphosing into two carbon dioxides will need two irons to discharge its electrons to. This makes the reaction chemically complicated as we have only one iron in direct contact with an oxalate ion. Therefore, iron(III) oxalates are stable compounds and will not, left on their own, decompose to iron(II) and carbon dioxide. They need additional energy input, and this is provided by light from the lamp in the enlarger,[¶] triggering the move of electrons from the oxalate to the two iron cations. Similarly, two iron(II) ions then need to meet one platinum 2+ ion to dump its extra electrons on, leaving microscopic metallic platinum(0) particles to form the image and returning the complex ion to its original iron(III) state. The iron oxalates are then washed away with water, often with some ethylene diamine tetraacetic acid, better known as EDTA, added, a ubiquitous and sometimes lifesaving molecule that we will hear more about in the next chapter.

So, look closely on the tag the next time you visit an art gallery or museum featuring new or old photographs, it may well be that the most attractive ones are platinum prints. In the words of a famous photographer of the American West, Laura Gilpin: "It's the most beautiful image one can get. It has the longest scale and one can get the greatest degree of contrast".[3] And, we might add, it does require the photographer to double as a chemist, as the papers are usually prepared from scratch (see Figure 3.2).

[¶] A projector for producing photographic prints from negatives by projecting the image onto a photosensitized paper mounted on a flat surface.

Figure 3.2   A photo of Craig Alan Huber's platinum print "Three Bells with a Voice, Mission San Miguel, Arcángel", reproduced with permission by the artist. On display in the Mission San Juan de Bautista museum, where in the bell tower, and by a complete coincidence, platinum blonde Madeline Elster/Judy Barton, in the shape of Kim Novak, meets her destiny in Alfred Hitchcock's Vertigo.

3.4   VERY, VERY RARE ELEMENTS

Gas light and photography are two great inventions of the 19th century, but the chemistry invented by Victorian author H. Rider Haggard in his 1886 bestselling novel *She* involves some almost alchemical transformations of the title character Ayesha, not to mention what goes on in the private laboratory the author has equipped her with. By coincidence, *SHE* is also the acronym of choice used by that modern breed of alchemists, the scientists who synthesise new *Super Heavy Elements* by smashing lighter atoms together and hoping for a fusion to a new element that eventually will take its place in the periodic table.

This, it may seem, is the exclusive playground for advanced experimental physics at large and very expensive "big science" facilities at the Joint Institute for Nuclear Research in Dubna, Russia, Lawrence Berkeley National Laboratory in Berkeley, California, the RIKEN Nishina Center for Accelerator Based Science in Japan, and the Institute of Schwerionenforschung (GSI) in Darmstadt, Germany. However, before the atom smashing game can begin, chemists have been at work.

One of the latest additions to the periodic table is element 117, confirmed by a joint IUPAC–IUPAP[‡‡] Working Party very late in 2015, and with the name tennessine and symbol Ts approved in 2016. The name is to commemorate, among others, the work of a very special group of chemists and physicists at the Oak Ridge National Laboratory (ORNL) in Tennessee, USA.

To make elements with a higher atomic number than 103, the so called superheavies, one ideally starts with as heavy starting materials as possible. The researchers will send a stream of atoms of one element bombarding a target of another element during weeks of painstaking experiments, hoping for that rare occurrence of two atom nuclei fusing into one. The continuous stream needs to be made of relatively abundant materials, but the target, a few thousandths of a gram only (it is typically about 12 to 15 milligrams), is a different story, the heavier the better.

And it turns out you can make just enough of some of the artificial elements in this way, for example californium and berkelium with atomic numbers 98 and 97 respectively. The process, however, is not for the fainthearted. First, a rather nasty mixture of mostly curium and americium obtained in a nuclear reactor is transferred into high security laboratory at Oak Ridge. These samples are the subjects of intense neutron irradiation to form the desired californium and berkelium isotopes. Further processing is done at the so called hot cells area of this lab, where no living creature has set foot since legendary nuclear chemist Glenn Seaborg sealed it in the 1960's. Once formed, berkelium

---

[‡‡]The International Union of Pure and Applied Chemistry, and the International Union of Pure and Applied Physics.

and californium isotopes immediately start to decay. Therefore the chemist would like to get their hands on the sample as quickly as possible. However, even nastier highly radioactive stuff is produced in the process, so the team needs to wait for the sample to "cool off" before it is allowed out of the high security lab. Once "cool" enough, the process of separating the many elements now present in the sample can begin, and just as with the REEs, oxalic acid and oxalates are used to grab and purify the desired elements.

The proud chemist can then hand over the 20 milligrams or so of metallic californium or berkelium to the physicists. There is only one slight glitch though, at Oak Ridge there are no facilities to smash them up with intense beams of other atoms, so this highly radioactive sample, decaying at the rate of more than 1% a week, needs to go for a ride.

Researchers in general are known to pass the most bizarre objects through airport security, but californium and berkelium are clearly over the limit even for us. So, when Krzysztof Rykaczewski at Oak Ridge was helping to organise the transport of 20 mg of berkelium-249 to Dubna in Russia, the sample was sent by air freight in a very big lead lined box.

Off it went, and the first hit on the tracking number showed the box turning up in Jamaica. This got Dr Rykaczewski flying through the roof, but a quick check with the freight company calmed him down, this was Jamaica, New York. Worse, however, was to come.

When the big box arrived in Moscow, some paperwork had been misplaced by the freighters. Uncompromising customs officials promptly ordered the box back on the returning flight to New York. And all the while the precious sample was losing berkelium atoms by the minute. Speedily provided with the missing documents, the box returned to Moscow only to be met with a "niet". "You should have faxed us first" was an impromptu developed new rule.

Same procedure again, and the clock was ticking. Third time around, the papers are OK, but a suspicious customs officer looks at the form stating the cargo to be 20 mg of berkelium, looks at the large box and then checks the paperwork again. "It says here you're importing 20 mg of berkelium, but this is a very big box. Obviously, you are smuggling something, will you please open it."

Not a good idea. But this time Russian customs could be convinced to be reasonable, seeing tiny vials, each containing 4 mg of highly radioactive ^{249}Bk in five large barrels inside the shipping box. The box arrived unmolested in Dubna with enough berkelium to start the experiment, and the rest, as they say, is chemistry and physics history (Figure 3.3).

3.5 GUILT BY ASSOCIATION

Finally, as we have seen the many practical and industrial uses of rhubarb, sorry, oxalic acid, in this chapter, we might ask if there is a point in making a distinction between "natural" and "chemical" ingredients in the things we eat and other everyday products, or if it is perhaps the properties of each molecule and compound that need to be considered, not its origin. In addition, the same molecule can be used for many things, being part of a biochemical process in our body does not stop it from being useful in a chemical process in a factory.

Putting it another way: would you really want to eat a vegetable that contains a chemical used to purify heavy metal ions, process photographs and that is used to produce metallic samples from the most notoriously radioactive mixtures on the planet? Even though this chemical has a rather easy-to-pronounce name?

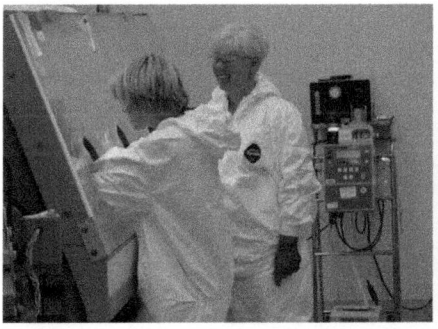 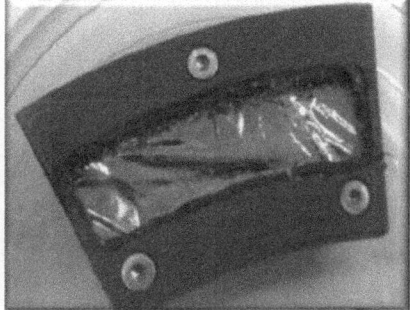

Figure 3.3 Left: ORNL chemists Drs Rose A. Boll and Shelley M. Van Cleve purifying a ^{249}Bk sample, photo courtesy of Krzysztof Rykaczewski, reproduced with permission. Right "One of the mixed-Cf target sectors electrodeposited on Ti foils at ORNL's REDC." Quotation and photo from Figure. 1 in K. P. Rykaczewski, *et al.*, EPJ Web of Conferences, 2016, **131**, 05005. Reproduced under the terms of the CC BY 4.0 license, http://creativecommons.org/licenses/by/4.0/.

We sure would! Get some rhubarb pie to celebrate the end of the chapter and then go on to learn about chelation therapy.

REFERENCES

1. G. B. Haxel, J. B. Hedrick and G. J. Orris, Rare Earth Elements—Critical Resources for High Technology, U.S. Geological Survey, 2005, Fact Sheet 087–02, https://pubs.usgs.gov/fs/2002/fs087-02/.
2. J. M. Ware, *J. Photogr. Sci.*, 1986, **34**, 165–177.
3. M. A. Sullivan, *Laura Gilpin*, Office of the State Historian, State Records Center & Archives, Santa Fe, New Mexico, http://newmexicohistory.org/people/laura-gilpin, read 8th September 2018.

CHAPTER 4

On Poisoning, Self-poisoning and the Euro

Meticulous sorting, sifting and collecting of our solid wastes, such as paper and metal, is perhaps something we associate with progressive thinking and sustainable development, but there is still an uphill struggle to reach the levels we need. Perhaps then, a sober reminder that we did it before is in place. In all countries affected by World War II, including neutral Sweden, materials recycling was an essential part of the war effort: paper, metal, and bones were collected, and you could even return surplus used cooking fat to your butcher (Figure 4.1).

The recycling taking place early in the war in Oxford barbershops was, however, a bit out of the ordinary. In all likelihood, it did not have the Oxford dons in false beards sweeping the floors and collecting cut-offs from young gentlemen getting their last proper trimming before joining the army, but the fallen hair did somehow find its way to a top-secret project in an Oxford university chemistry laboratory.

The Rhubarb Connection and Other Revelations: The Everyday World of Metal Ions
By Lars Öhrström and Jacques Covès
© Lars Öhrström and Jacques Covès 2019
Published by the Royal Society of Chemistry, www.rsc.org

Figure 4.1   Recycling was crucial on the home front in World War II, even
for neutral countries. Paper, metal, bones, and even surplus used
cooking fat were collected.

4.1   NICE-SMELLING DEW OF DEATH

When war with Germany started to seem inevitable to the UK
in 1939, a great fear was that the powerful German chemical
industry would be able to supply the Nazi army with large
quantities of chemical weapons. One of the major concerns
was an arsenic-containing compound called Lewisite, which,
because of its geranium-like smell,[†] was nicknamed Dew of
Death (Figure 4.2).

Ironically on many levels, lewisite was first synthesised by the
catholic priest, and later Professor of organic chemistry at the
University of Notre Dame[‡] in Indiana, Julius Nieuwland. Lewisite
contains a carbon–carbon double bond, and eventually Nieuwland
would go on to lay the foundations of synthetic rubber, large

---

[†]We have not tested this ourselves so we cannot be sure, but most likely we are talking
about the genus *Pelargonium* commonly known as geraniums, whereas the genus
Geranium, confusingly, are slightly different plants.

[‡]Where, we suspect, he is chiefly remembered for having had as a student and laboratory
assistant, the Norwegian Knut Rockne, who then left chemistry to become legendary
American football coach for the Notre Dame *Fighting Irish*.

lewisite

Figure 4.2 Lewisite, "Dew of Death", otherwise known as 2-chloroetheny-larsonous dichloride. The molecule has a central arsenic 3+ ion (grey-black, the metal), two chloride anions (light grey) attached to the metal ion and a HC=CHCl organic unit bonded to the metal with an As–C bond, thus making this an organometallic compound.

polymers containing many such double bonds.[§] Apart from being hospitalized for a few days after the synthesis, Nieuwland had nothing more to do with this molecule, but his thesis was picked up 14 years later by a certain Winford Lee Lewis. He was director of a Chemical Warfare Service unit in Washington, D.C. and found that the compound was just the ticket he needed. As the war ended, the US has stockpiled about 150 tons of the substance, but fortunately never used a single dose in battle, meaning that chemical warfare could largely be pinned on the losing German side.

But now, in 1939, UK concern was that the Germans were going to use lewisite in the coming war (the Mussolini regime in Italy had already used it in Ethiopia), and a counter measure was needed. Evidence had accumulated during the inter-war years that arsenic compounds attacked vital sulfur–hydrogen bonds in some key enzymes (proteins that catalyse chemical reactions in the body) responsible for the vital conversion of glucose to usable chemical energy in the body (the molecules ATP and NADH). To test possible antidotes, the Oxford team (Lloyd Stocken, Robert Thompson and others) needed a protein that could somehow mimic these human enzymes that had not yet been isolated. Owing to the war, it was even harder to obtain well characterised non-human enzymes, so the choice fell on *keratin*, a protein packed with the S–H containing amino acid cysteine.

[§]As shown for oxalic acid in the preceding chapter, a molecule often has multiple uses, and this educational website from the Organisation for the Prohibition of Chemical Weapons (Nobel Peace Prize in 2013) explores the beneficial uses, misuses, and abuses of multi-use chemicals, both historically and presently. https://www.opcw.org/special-sections/education/multiple-uses-of-chemicals/.

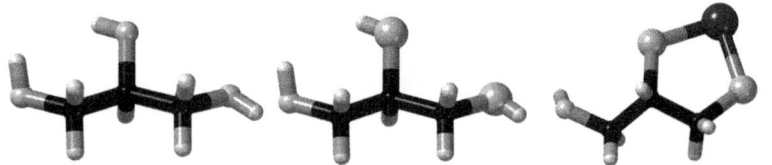

Figure 4.3   Left: glycerol, HOCH$_2$CHOHCH$_2$OH, centre: British Anti Lewisite (BAL), HSCH$_2$CHSHCH$_2$OH and right a chelate formed by arsenic 3+ and BAL.

Keratin is a major component of hair, and indeed the same keratin that, somewhat recklessly perhaps, figures in discussions and advertisements about curly hair. The key to frizzy or un-frizzy hair is the sulfur–sulfur bond that exists between some cysteines in different keratin helices, and we will encounter similar bonds later in this chapter. They can be broken by reduction and reformed by oxidation and in-between the hair can be more or less freely formed. Both processes require potent chemicals, so take care if you are inclined to test these things out. The attraction for the Oxford boffins was that this protein was readily available on the floor of any barbershop. So off they went collecting and could soon start the experiments that led to a compound that would be appropriately named British anti-Lewisite once it crossed to the other side of the Atlantic (Figure 4.3).

The essence of the keratin work was to show that the lewisite delivered the arsenic ion to the protein where it would bind to two cysteine S–H units very close to each other. The logical antidote was then to find a small molecule that would bind better to the arsenic, and be water soluble, so that instead of being sick the soldiers would just pee out the broken-down arsenic-containing chemicals of the lewisite. The final choice, that Stocken also tested on himself, was a compound very close to glycerol, a small tri-alcohol obtained from fat, but in which two of the –OH groups, the very definition of an alcohol, had been substituted for two –SH groups, just as those in cysteine.

In the following tests, British anti-Lewisite was found to be very effective, and was produced on a large scale both in the US and the UK and distributed to the armed forces. It was never used, however, as lewisite stayed in the storage tanks on both sides during the war.

4.2   DOCTOR'S ORDERS: CHELATION THERAPY

When the secret veil was lifted from the lab benches after the war, one obvious, completely civilian, application quickly appeared, the use of British anti-Lewisite against arsenic poisoning in general. And not only that, once BAL had proven its efficiency against arsenic, and later mercury, a whole new way of curing metal poisoning was created: *chelation therapy* based on similar compounds that would encircle the offending metal ion, form a soluble compound, and then disappear from the body in the urine taking the metal ion with it.

Care must be taken though, so that the metal ions do not end up in a more damaging place than before. So, for example if BAL is used against thallium or cadmium, other big and soft (remember the silver polish) metal ions that will "love" this sulfur ligand, the so formed molecules may end up causing more damage in the kidney or the brain than if one just lets them slowly seep out.[1]

Another concern is the use of chelating molecules in so called "alternative medicine". Efficient as they are against metal poisoning, when taken without any such indications they will do no good and may more often than not be harmful. If nothing else, they might transport essential metal ions such as zinc, iron and copper out of your body.

4.3   METAL MANAGEMENT

When London neuropathologist John Nathaniel Cumings analysed copper in diseased and healthy brains in 1948, he saw high concentrations in some cases, not related to any known accidental copper poisoning of the former owners of the organs. Copper is an essential element that we need for many things, but these abnormal concentrations were potentially problematic, and suggested that the bodies poisoned themselves through some kind of copper mismanagement. The idea then occurred to him that, perhaps an internally generated poison could also be treated with BAL?

He went on to test this and only three years later, provided the first drug against Wilson's disease. This is a genetic condition that, if left untreated, will eventually be fatal, causing liver and neurological damage when the body is not capable of controlling the copper concentrations in the liver.

Better treatments were eventually found, BAL is not an innocent molecule, but most drugs against Wilson's disease are still based on chelates.

Such developments probably did not go unnoticed in the "communist block", even though information did not flow as freely then as it does now. Or, as was known to happen, they had a few good ideas of their own. In any case, in the 1950's Russian chemists developed a close relative of BAL having the same arrangement of the sulfur atoms, but now with carboxylic acids on both ends. This molecule, known simply as DMSA,[¶] was extensively researched and also used in China and Japan, but it took until the mid 1970's before the news of its efficiency against lead and mercury poisoning spread to the West. One can but speculate why. Chinese chemistry might have been looked down upon during this period, and Chinese researchers had a difficult time throughout the Culture Revolution, but Russian science was closely followed by the West,[‖] and all major journals were translated into English.

In any case, DMSA is now on the World Health Organization's list of essential medicines, and was an important tool in rescuing hundreds of children in Nigeria's northern state Zamfara in 2010. During routine meningitis surveillance at the beginning of the year, local public health officials and *Médecins Sans Frontières* (MSF) came across a disturbing high number of small children suffering convulsions and spasms. These might have been caused by meningitis infections or malaria, but as both industrial and "illegal" artisanal gold-mining had recently taken off in a big way in the region, metal poisoning was also a possibility. The physicians knew lead poisoning typically affects the brain and could have caused the convulsion symptoms. And indeed, no infections were found, but many children had such high lead levels in their blood that they were life threatening.

[¶]Short for di-mercapto-succinic acid, in which succinic acid is $HOOCCH_2CH_2COOH$, once known as spirit of amber and a common chemical in industry, and also naturally occurring in our bodies.

[‖]As a young engineering student in Stockholm in the mid 1980's one, of the authors remembers being told in a compendium on chemical literature that "learning Russian was not such a bad idea for a chemist".

It was immediately clear that this was potentially grave, and that many villages might be suffering. The grinding and processing of ores sometimes took place inside houses, and with equipment also used in food preparation. To make things worse, this ore contained very high amounts of soluble lead compounds, and moreover, as the activity was illegal, figuring out which places might be stricken could be tricky.

It was an outbreak, or a kind of epidemic if you want, and it was lethal. The Nigerian Federal Ministry of Health assembled a multidisciplinary task force with both national and international (WHO, MSF and the United States, Centers for Disease Control and Prevention, CDC) experts. Among many actions taken, DMSA, that can be taken orally and does not need to be injected, was provided on a large scale. This probably saved hundreds of lives.

We don't know how the ore was processed, but a popular but dangerous method, unfortunately still practiced a lot in illegal and artisanal goldmining, is to draw the gold out from the crushed ore using metallic mercury (a liquid at room temperature). It is kind of ingenuous, as the gold will dissolve in the mercury and the rest of the stones and debris can then easily be disposed of. The mercury can subsequently be evaporated away over an open fire and the happy miner is left with solid gold but also a grave risk of mercury poisoning. Be that as it may in this case, DMSA treatment would also have taken care of the mercury.

4.4 THE FLIP-FLOPPING MOLECULE

We cannot continue this chapter without talking about 2, 2′, 2″, 2‴-(ethane-1,2-diyldinitrilo)-tetraacetic acid, better known as EDTA, short for ethylene diamine tetraacetic acid.** A mouthful, indeed, but the standard tool for picking up unwanted metal ions wherever they may be, in your water or in your body, in the factory

---

**The common name can be decoded as follows: "ethylene" is $H_2C=CH_2$ and we can possibly attach one more atom (or group) to each carbon by breaking the double bond. These are the two amine (NH_2) groups in "diamine". In the NH_2 we can then change the hydrogens for something else and if we do this with $-CH_2COOH$ on both nitrogens we get the "tetraacetic acid" part. It has been named one of the 100 most important chemicals.

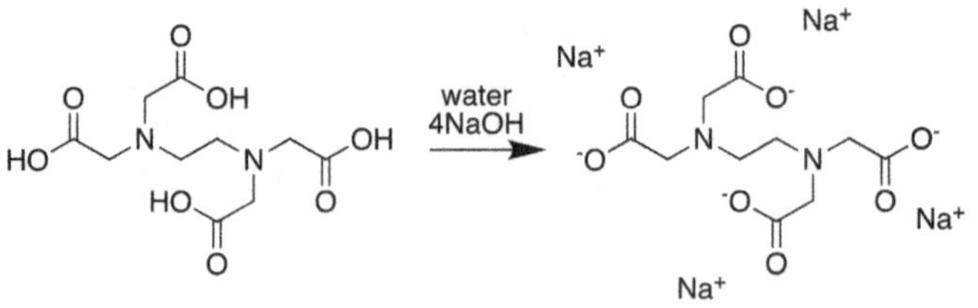

Figure 4.4   H$_4$EDTA and how it becomes the efficient six-dented Na$_4$EDTA.[††]

or in the chemist's round-bottom flask in the lab. Therefore, here comes a chemical formula and a reaction (Figure 4.4).

First prepared by German industrial chemist Ferdinand Münz in 1936, as he tried to get rid of problematic metal ions while dying textiles, this is a completely synthetic molecule and is not found in nature as far as we know. Münz was successful in improving his colouring process, and in doing so helped the German Nazi regime to be less reliant on chemical imports such as citric acid, the previously used agent. This did not prevent him from being deported to the Theresienstadt concentration camp during the war. He survived, however, and could then follow the worldwide success of his floppy molecule.

For floppy it is, and we need to dwell on this for a while.

If we draw a picture of a lady on a bicycle, it is fairly obvious to most of us that if the said lady's feet are not grounded, the bike needs to be moving, legs pumping up and down, and the wheels are spinning. In other words, what we see in the picture is an over-simplification, we use a static representation of something that is really very dynamic, and constantly moving. It only works because most of us know how to ride a bike. We unconsciously add some prior knowledge and experience when our brain interprets the picture.

The prior knowledge we need to have when it comes to mole-cules is that these too are constantly moving. A real molecule is nothing like the stiff ball-and-sticks model you might have used in school. Instead, you need to imagine the EDTA-molecule as having loosely fitted and freely moving joints between all sin-gle-bonded atoms. A slight push from a colliding molecule will send one part of the EDTA spinning around a C–C axis, adopting

---

[††]Actually, as the amine part is a base and the acid an acid, two of the protons will sit on the nitrogen atoms not on the oxygens.

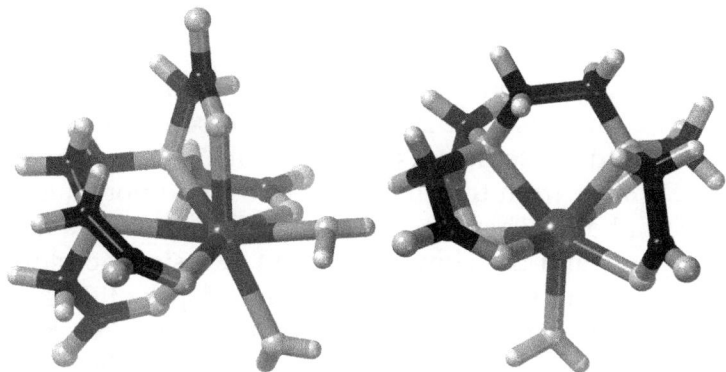

Figure 4.5   Left: Taking care of hard water, an EDTA 4− ion wraps itself around a calcium 2+ ion efficiently masking it from interacting with anything else, such as carbonate ions that could make an ugly scale of calcium carbonate in unwanted places. Right: An EDTA 4− ion wraps itself around a manganese 2+ ion in the process of bleaching paper pulp thereby hindering Mn^{2+} ions from the wood from catalysing the breakdown of the environmental friendly bleaching agent hydrogen peroxide. In both molecules additional water molecules bind to the metal.

a new conformation. A bit like a jewellery chain left on a flat surface that easily can be pushed into different forms. Once the chain is around the neck of an upright person, it is a different story. Basically, only one conformation is now possible because the chain is threaded around a neck and gravity pulls it downwards into a nice curved shape.

The same happens to the flip-floppy EDTA molecule. The four arms wave around in all directions, but once they encounter a metal ion they will lock on to it, and one by one the other arms and the central part with the two nitrogen atoms will follow, and soon the metal ion is all wrapped up by the ligand, efficiently sheltering it from the surroundings (Figure 4.5).

Another thing we need to know about the way we draw pictures of molecules is that pictures such as those above are simplified in another way. Drawing a molecule in this way is like wanting to draw a human, but all you can manage is an outline of the skeleton. Which is good, if you want to see how hands are connected to arms and legs to the body, and how they can move and how the joints are fitted. What we are doing with a traditional ball-and-stick model is then to emphasise the same things, chemical bonds and how atoms and groups can move relative to one another.

But, if we want to know if this human is thin or fat, a body builder or a long distance runner, we need some more meat on the bones. And the same goes for molecules, the electrons move around far away from the chemical bonds, and not all are even tied up in a bond. There are various ways we can measure and calculate this, and the simplified result is a specific radius for each element and this is what we use in a "space filling" drawing. These try to capture the volume of a molecule, such as that below for the manganese complex with EDTA shown in the picture (Figure 4.6).

4.5   ANOTHER EMPEROR, HIS HORSE AND HIS TAPESTRY

Metals may also be part of vital drugs, as we have already seen, but these may in their turn give adverse effects. There is no point in curing the patient from one disease only to kill him or her with the drug. Therefore, one of the first uses of British anti-Lewisite, once it was clear it was not going to be needed against German shells, was against voluntary arsenic poisoning from the popular neosalvarsan drug, the only efficient treatment for syphilis until the arrival of penicillin after the war.

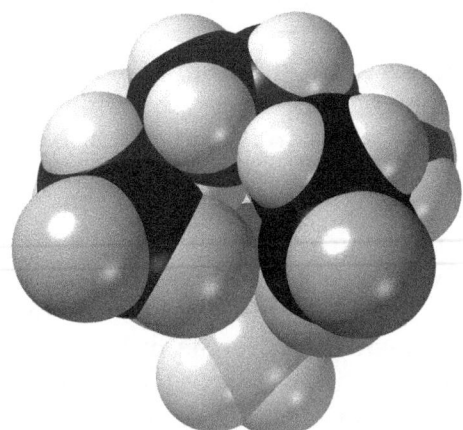

Figure 4.6   A so-called space-filling drawing of the Mn(II)EDTA complex in the preceding figure. Here there is more meat on the bones.

This invention was, however, too late for Napoleon Bonaparte, and others in the 19th century believed to have been poisoned by arsenic fuming off from wallpaper coloured by Scheele's Green, $CuHAsO_3$, the most popular green pigment of the 19th century. The death of Napoleon still engages researchers, and while the wallpaper in Longwood on Saint Helena indeed seems to have contained arsenic, the cause of his death at the early age of 51 is still contested.

The case of Napoleon's horse Nickel may therefore be less contentious, although even more folkloric. On his way to the decisive Austerlitz victory in 1805, Napoleon was riding horses from the Imperial stud farm in St Cloud, among which was a white-grey horse named Nickel because of its colour. Stopping and asking for bread, perhaps hoping for a crunchy French baguette, the disappointed emperor was instead served dark Westphalian rye bread. According to the folktale, Napoleon was not amused, and declared the bread not suitable for him-self, the Emperor, but better fit for Nickel, his horse: "*C'est juste bon pour Nickel!*" which evolved from "bon pour nickel" into pumpernickel.[‡‡]

The Napoleonic efforts to unify continental Europe were eventually thwarted, even though the Code Napoleon, replac-ing many feudal and medieval laws in "liberated" Europe, remained in place. Not a fan of democracy, except for getting himself elected, Napoleon might nevertheless have approved of the 1992 Maastricht Treaty that established the Economic and Monetary Union and laid the foundations for the single cur-rency. Ten years later, on 1 January 2002, euro banknotes and coins were put into circulation, giving us another reason to talk about nickel.

A characteristic of the one and two euro coins is that they are made of two different alloys. In the one euro coin, the yellow ring is made of copper with 20% zinc and 5% nickel (nickel brass) while the inner component, the white pill, is cupro-nickel: cop-per with 25% nickel by weight. It is the opposite with the two

[‡‡]The etymology of pumpernickel is unclear, one suggestion is that is has to do with a German variation of "Old Nick" for the Devil or an evil spirit, which gives the bread a shared origin with the chemical element nickel whose name has its roots in Kupfernickel, the mineral that unfortunately, or because of evil spirits, is not copper despite the appearance.

euro coin, with its white ring and its yellow pill. When the metal gets a bit tarnished, the bimetallic composition effectively turns the coin into a small short-circuited battery, promoting galvanic corrosion, and nickel release, especially when the coins are in long contact with the skin and the sweat. This was measured during a study conducted by the Institute of Metallurgy of Zürich, Switzerland, and after prolonged contact with human skin, the nickel release from the one and two euro coins exceeds, by a factor of 240 to 320, the values acceptable by the European Union Directives.

A direct consequence of nickel release in contact with the skin is the development of contact allergies.[§§] Up to 5–13% of the population are allergic to this metal and can suffer from severe dermatitis giving itchiness, red skin, and a rash. Urinary excretion of nickel contributes to improvement of dermatitis such as hand eczema and can be enhanced by chelation therapy for example with sodium diethyldithiocarbamate (DDC), a substance used for the treatment of nickel carbonyl poisoning, the effects of a rather odd molecule with the formula $[Ni(CO)_4]$ (Figure 4.7).

Two molecules of DDC contribute to nickel binding to form a compound with the metal ion in the centre of the square with sulfur atoms in the corners, which we call a square planar

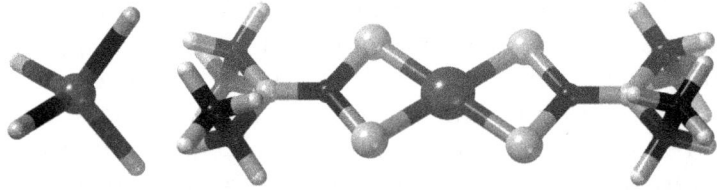

Figure 4.7   Left: The poisonous tetracarbonylnickel, also known as nickel tetracarbonyl, molecule, a neutral nickel atom with four carbon monoxide molecules attached to give $[Ni(CO)_4]$, probably one of the most toxic substances routinely used in a large scale by industry. Right: Two diethyldithiocarbamate (DDC) ions, used to treat tetracarbonylnickel poisoning, with an incapacitated nickel 2+ cation in-between.

[§§]Perhaps surprisingly, massive coin ingestion is another problem, especially for children aged 1 to 5 years, being the main foreign bodies ingested (27%), far ahead of batteries (13%) or toy parts (12%). Anyway, nickel in excess in the diet is an aggravating factor for nickel-hypersensitive patients.

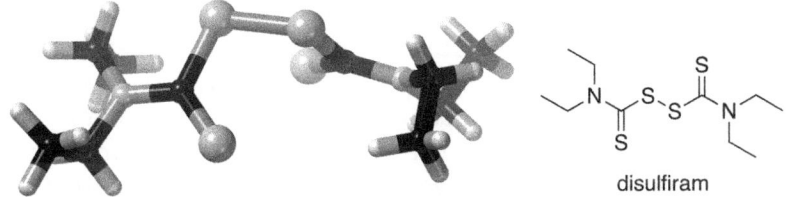

disulfiram

Figure 4.8   Tetraethylthiuramdisulfide (TETD), also known as disulfiram or antabuse.

geometry. This molecule will easily be eliminated by the urinary tract, and the efficacy of the dithiocarb treatment was evidenced when four workmen working on the manufacture of acrylic plastics were accidentally exposed to vapours of nickel carbonyls and soon developed severe respiratory symptoms (reported in 1979). Three of them received dithiocarb medication from the industrial dispensary, and recovered within 48 hours, while the fourth poor guy, who instead consulted his family physician, was given an injection of penicillin instead of dithiocarb administration and died four days later.

4.6   A COCKTAIL CONNECTION

DDC used to treat nickel poisoning is also the product of the metabolism of another molecule with a scary name, tetraethylthiuramdisulfide (TETD), also known as disulfiram or antabuse, a drug used to fight alcohol addiction (Figure 4.8). Disulfiram consists of two molecules of DDC linked by a disulfide bridge (remember the curly hair) and is split into two molecules of DDC by a simple reduction, just as when you want your hair to behave. This explains why disulfiram can be used to treat nickel dermatitis as well. However, some surprises can arise.

Ethanol, or for simplicity alcohol, is broken down in two steps. First an oxidation will produce acetaldehyde. This is a compound widely found in nature, but nevertheless both an irritant and probably also a carcinogen. So, to take care of that, another oxidation step is needed to produce harmless acetic acid. As disulfiram stops the enzyme that does this, *acetaldehyde dehydrogenase*, (meaning it will help to take a hydrogen away from a carbon in acetaldehyde and C–**H** will become C–**O**)

while the enzyme doing the first oxidation just keeps on working.¶¶ This means that acetaldehyde will accumulate, first in
the liver, and then in the blood. Once the acetaldehyde concentration in the blood starts to increase, the hangover symptoms appear directly, not the day after, going from nausea,
vomiting and throbbing headache and even circulatory collapse. This supposedly will discourage you from any alcohol
absorption.

In addition to these unpleasant side effects, one patient
treated with disulfiram for alcohol addiction developed a puz-
zling acutely intense pruritic rash on his left wrist, just as for a
nickel allergy. It turned out that 15 years earlier, this person had
been diagnosed as nickel sensitive, having developed a contact
dermatitis caused by nickel in his watchstrap. Similar reactions
were reported for persons sensitive to rubber or cobalt and put
in contact with disulfiram or its derivatives. These flare ups,
often more severe than regular dermatitis, are due to hidden
nickel deposits in the body, suddenly dragged out by reaction
with the DDC that has been produced by the reduction of the
disulfiram and giving rapidly rising serum levels of nickel. Any
history of nickel or rubber contact dermatitis must therefore be
explored before disulfiram therapy is attempted in an alcoholic
patient.

The connection with rubber is due to the fact that zinc complexes of disulfiram or of similar compounds are used as vulca-
nizing agents for rubber tires. It was observed that workers in
these industries had very bad reactions to alcohol. This observation remained unexplained until Erik Jacobsen, with Jens Hald
and Keneth Ferguson at the Danish drug company Medicinalco,
demonstrated the effect of disulfiram and alcohol on themselves
during a self-experimentation session.

Mentioned less frequently than ethanol in literature, disulfiram nevertheless appears in *Miss Smilla's Feeling for Snow*
(Peter Høeg, 1992) where a certain Julianne has some difficulties waking up after consumption of both alcohol and the
drug.

¶¶This one we call *alcohol dehydrogenase*, as it will take away the first C–H hydrogen from
ethanol to give a C=O.

4.7 PARASITE POISON

Youyou Tu and the Běncǎo Gāngmù are not world-famous web video stars who top the latest music charts. Instead one is the 2015 Nobel Prize laureate in Physiology and Medicine, and the other the most comprehensive and understandable book to deal with traditional Chinese medicine (written by Li Shizhen between 1552 and 1587, Ming Dynasty), respectively. Youyou Tu was awarded the Nobel Prize for her discoveries concerning a novel therapy against Malaria using artemisinin, a substance which inhibits the malaria parasite, extracted from *Artemisia annua* (sweet wormwood), a plant described several centuries ago by, among others, the said Li Shizhen.

There is nothing fundamentally new in this approach, looking into traditional remedies that have their sources in nature. Chemists are on the constant lookout for biologically active compounds from plants and other living sources and have been ever since the start of the discipline. We will encounter a few other examples later on. One aspect of interest in the story of Youyou Tu and artemisinin is how this drug was developed in the midst of the Chinese Cultural revolution and the Vietnam war (during which malaria was a big issue for both sides), and eventually found its way also to "the West". But we are digressing into the relationships between science, innovation and society, a passionately important subject, but one that requires its own book.

Antimalarial treatments essentially involve two combined strategies: to fight the parasite *Plasmodium* spp., with *P. falciparum* as the most virulent strain, and to target the second phase of the disease, as two phases characterize the malaria infection cycle. When a person is infected after a blood meal by an infected mosquito, the parasite travels first to the liver for a long phase during which the victim has no symptoms, but during this phase the parasite yields thousands of particles (the hepatocytic phase). These parasite particles eventually escape from the liver cells, spread into the blood flow and infect the red blood cells, this is therefore known as the erythrocytic phase. The multiplication and survival of the parasite in the erythrocytes (red blood cells) depends on its capability to destroy the haemoglobin, in order to use the resulting amino acids for its own development. This behaviour is responsible for the cyclic periods of coldness and shivering followed by strong fevers and sweating, and the severe anaemia which is characteristic of this disease.

We have seen earlier in this book (Figure 1.6) that haemoglobin contains a specific iron chelator called the heme group. The iron ion in this molecule is used for the reversible fixation of dioxygen. In other words, haemoglobin catches dioxygen from the air coming into the lungs and transports it to the organs through the blood flow. This is the physiological definition of respiration and the way in which we breathe. When the parasite destroys the proteic part of haemoglobin for its own profit, it also releases an iron-containing compound called ferriprotoporphyrin IX that is in turn toxic to the parasite. The metal is pretty tightly bound to the porphyrin unit, so it is not by all means clear that chelation therapy would help the parasite. Therefore, it has developed its own completely different strategy for dealing with this toxic compound. The parasite prevents the deleterious effects of ferriprotoporphyrin IX by adding the individual molecules together to form a compound with an insoluble crystalline form called hemozoin, or malaria pigment (Figure 4.9).

Survival of the *Plasmodium* relies on hemozoin formation. This step is therefore an attractive target for antimalarial drugs. The first used were quinine derivatives, including chloroquine, and then came the second generation of artemisinin drugs.

The mechanisms of action of these compounds are not fully understood so far, but they probably use a chemistry similar to that known for producing reactive oxygen species, or ROS, and which depends on the redox state of the iron ion.

Iron in haemoglobin is Fe 2+ and can turn to Fe 3+ when heme is released from the protein scaffold or when iron is extracted from heme. This reaction is an oxidation, meaning that Fe 2+ is losing an electron. However, this electron needs to go somewhere, and it can be taken up by hydrogen peroxide, a natural oxidant that will be normally found in small amounts in a red blood cell. The problems (for the parasite) start with the production of one of the ROS: the hydroxyl radical ·OH (this first

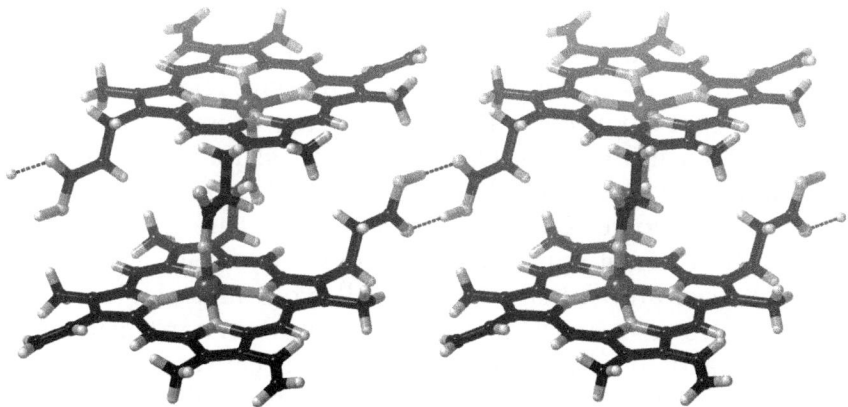

Figure 4.9 The malaria parasite *Plasmodium falciparum* does not have access to chelation therapy and it is unclear if this would help it anyway. It avoids iron poisoning by making long insoluble hydrogen-bonded chains of the haemoglobin remains. This compound is called hemozoin, or malaria pigment, because of its black-brown colour that can be observed in the blood of malaria sufferers.

reaction is known as the Fenton reaction). Then, Fe 3+ can be reduced back by another ROS present in the red blood cell, the superoxide anion $\cdot O_2^-$, itself being oxidized to dioxygen. In the net reaction, iron has a catalytic role, making the process even more lethal for the parasite.

$$\cdot O_2^- + H_2O_2 \rightarrow \cdot OH + OH^- + O_2$$

This reaction is known as the Haber–Weiss reaction after the notorious German chemist Fritz Haber and one of his students Joshua Weiss. The manuscript describing this reaction was submitted by Fritz Haber a few days before he died in 1934. Fritz Haber will feature a few times in this book, because he had multiple successful, but also repugnant, activities. For instance, he was both the 1918 Nobel Prize laureate (received in 1919), for the innovation of the catalytic production of ammonia from hydrogen and atmospheric nitrogen under high temperature and pressure (the Haber–Bosch process) and a recognized Word War I criminal for his deep involvement in the development and personal release of different deadly gases (from chlorine to Zyklon A, and then B).

During its red blood cell reproduction cycle, *Plasmodium falciparum* causes the disintegration (lysis) of 80% of its host cell's haemoglobin molecules, which gives a massive release of pro-oxidant ferriprotoporphyrin IX catalysing the production of ROS, the most harmful being ˙OH. The hydroxyl radical cannot be eliminated by an enzymatic reaction, which is possible for the superoxide anion or the hydrogen peroxide.[ll] Thus, even if it has a very short life of approximately one nanosecond, it will have time to react and damage virtually all types of biological macromolecules such as DNA, RNA, lipids, proteins or sugars. The unique reactivity of this molecule probably causes the death of *Plasmodium* when the hemozoin production is inhibited by artemisinin and diverse quinine derivatives.

This might have been a bit more complicated than you bargained for when you bought the book, but malaria is a complicated problem. We will return to some less intricate chemistry in the next chapter, with more chelation therapy and learn why gold production is potentially so poisonous. And we will do so in the company of queens, conmen and criminals.

REFERENCE

1. M. W. Shannon, S. W. Borron and M. J. Burns, *Haddad and Winchester's Clinical Management of Poisoning and Drug Overdose*, Saunders, Philadelphia, US, 4th edn, 2007.

[ll]The superoxide anion is taken care of by superoxide dismutase, and hydrogen peroxide by catalase enzymes.

Down to Earth Chemistry

Clothes and jewellery, nylon stockings and gold earrings, adiponitrile and gold 1+ ions, as we look closer and closer at everyday things they get more and more chemical, and unexpected connections turn up. In this case cyanide, a chemical base for both nylon stockings and gold earrings. We are, of course, in this book more interested in the gold earrings, although a fair number of metal ions are also involved as catalysts in the industrial manufacture of nylon.

5.1 DOWN UNDER WITH CHEMISTRY

A remarkable discovery was made in the late 19th century: the sea was full of gold. Or, at least if you took the tiny, tiny concentrations, you needed 100 tons of seawater to get 1 gram of gold and multiplied with the unfathomable volume of the world's oceans, you ended up with a theoretical amount that was quite astonishing. And more importantly, even in those high days of robber capitalism, nobody could charge you for the seawater.

It was soon clear though that knowing it was there was one thing, getting your hands on it was quite a different story. But the lure of quick and easy money was of course irresistible for some.

The Rhubarb Connection and Other Revelations: The Everyday World of Metal Ions
By Lars Öhrström and Jacques Covès
© Lars Öhrström and Jacques Covès 2019
Published by the Royal Society of Chemistry, www.rsc.org

Whether the Electrolytic Marine Salts Company at North Lubec, Maine, was a hoax from the beginning, after all the founder Prescott Ford Jernegan was a Baptist minister, or if it started in earnest, we do not know. But by the time innocent New Englanders were flocking to buy shares in the company, all the gold harvested in the mercury accumulators in the former tide water mill came from discharged jewellery.

This rather well known 1898 fraud may have made investors more suspicious, but nevertheless, in 1904 the UK company Industrial and Engineering Trust Limited of Henry James Snell attracted a number of well-known investors among its shareholders. Perhaps because the recent Nobel laureate in Chemistry, Sir William Ramsay, vouched for the project, a fact his biographers discretely by-pass. Because by 1905 this company, although by all evidence an earnest attempt, had also disappeared.

Both these projects were based on the sorption of gold into liquid mercury to make an amalgam, a solid solution of two or more different metals, and the subsequent evaporation of mercury simply by heating, leaving solid gold behind. Based on inexpensive mercury this was both cheap and simple but also dangerous for workers and for the environment. And not much chemistry is involved, so it is also boring.[†]

Not so for the Broken Head project on the northern coast of New South Wales.[‡] Here, also in 1904, Australian mining engineer Alfred Argles used his own money to invest in a scheme that had two inventions. It used a wave-powered motor to drive all pumping and other mechanical devices in the plant, and it used, as its final step, cyanide ions to separate the gold from the other metal ions.

This process was relatively new but had been successfully used in Australia for more than a decade. This chemical separation of gold from the less noble components of the ore was both more efficient and gave higher yields that trying to fish out the miniscule gold particles themselves. It is based on the concept that gold ions are big and soft. So, they will form bonds with other soft compounds or ions, and one such, also very cheap ion,

---

[†]This was for a long time the standard method of extracting gold from low grade ores, but nowadays it is mostly practised in legal or illegal small artisanal gold mining, where of course it is even more difficult to contain the dangers than in a proper plant.
[‡]Between Cape Byron and Lennox Head, lat. 28° 44′ S, long. 153° 37′ E.

Figure 5.1   The dicyanoaureate ion, [Au(CN)$_2$]$^-$, is the result when cyanide
ions (CN$^-$) and oxygen are used to dissolve gold to give gold +1
ions. In this way gold can be separated from other components in
the ore and then purified.

is the cyanide ion. Together they form compounds with a gold
atom at its centre and two cyanides forming a rod-like ion
(Figure 5.1). And again, we might add, it is based on something
extremely toxic. Cyanide ions are as iconic in detective stories as
arsenic.

But how do you get the gold 1+ ions from the metallic gold in
the first place? You might have learnt that the only thing that
dissolves gold is *aqua regia*, a mixture of concentrated nitric acid
and concentrated hydrochloric acid. It is a good method if you
need to hide solid gold, as when George de Hevesy dissolved the
Nobel Prize medals of Max von Laue and James Franck just hours
before occupying Nazi German forces stormed into the Bohr
Institute of Theoretical Physics in Copenhagen. De Hevesy himself eventually fled to Sweden, but the anonymous yellow-orange
tetrachloroaurate ions, [AuCl$_4$]$^-$, swam around undisturbed in a
glass container high upon a shelf for the rest of the war. Back in
Copenhagen after the German capitulation he reversed the reaction and sent the gold back to Stockholm where it was recast into
new medals, presented to Laue and Franck in 1952.

But *aqua regia* dissolves most things, and is not a practical
alternative for gold refining, and this is where the cyanide ions
come in. As cyanide ions form very strong bonds with gold, its
reluctance to give away its electrons, thus its noble character, is
lost. The lion part of all gold produced today has gone through
such a process. Leached by a cyanide solution the gold metal
in the tiny deposits becomes vulnerable to reactions with oxygen, and water soluble [Au(CN)$_2$]$^-$ ions are formed as oxygen is
reduced to water, just as when we breath. When the dicyanoaureate[§] ions, as they are called, have been separated from the

---

[§]Funny name, but when these metal ions form large more complex ions with an overall
negative charge instead of a positive, we no longer name these species after their English
name but after the name of the metal in Latin. In this case aurum, as in the symbol Au.

remaining particles we can get the gold metal back again by adding zinc metal for example, a very un-noble metal that will be more than happy to shed two electrons to give one Zn^{2+} that will swim out into the solution, and two atoms of metallic gold that will fall down to the bottom of the vessel.

Prescott Jernegan supposedly made a few bucks before fleeing the field, Henry James Snell did not, but walked away still an honest man, but the fate of Alfred Argles' gold-sequestering factory is not really known. It was embedded on a coast where titanium and zirconium sand-mining would later be profitable, and the remains have also been mistakenly attributed to these later operations. Brett J. Stubbs, writing in the scholarly journal Australasian Historical Archaeology in 2008, speculates that the foul weather on the coast might have brought the dreams to an end before it was put to the test. "There's many a hope been borne out to sea on a rough night on the northern beaches", as one old miner on the coast put it in the 1930's.

We can only hope Mr Argles was well insured and invested whatever he could claim more wisely, as it seems the concentrations of gold in seawater was greatly overestimated in the late 19th century, and accurate measurements were not made until the 1990's.[¶] The notorious German chemist Fritz Haber of gas-war fame thought he could pay the German World War I debt single-handed by extracting 6.5 g of gold from every 100 metric tons of seawater using his own process based on electrochemistry and centrifugation. However, Haber made some initial calculation errors leading to an overestimation of a factor of one-thousandth of the actual amount of precious metal, and his later measurements of the gold concentrations gave such low figures that he realised the process would never be profitable. A pity, as the Second World War might perhaps have been less likely had the huge debt imposed by the Allies in the Versailles Treaty been easier to pay.[‖]

[¶]Modern measurement puts this figure at 50 femto-mol per litre in the Pacific and the Atlantic, with the Mediterranean deep waters containing higher concentrations in the range 100–150 femto-mol/l (femto is 10^{-15}). K. Kenison Falkner and J. M. Edmond, Gold in Seawater, *Earth Planet. Sci. Lett.*, 1990, **98**, 208–221.

[‖]One could argue that Haber payed off his debt for the war time horrors of the gas-war by inventing the Haber–Bosch process, enabling us to harvest nitrogen atoms from the atmosphere and keeping roughly 60% of us alive today.

5.2 THE GOLDEN STOOL AND THE CYANIDE CODE

Gold is probably the most iconic element, known by the ancients, and perhaps manifested best, both in terms of the past and the present, by the Ashanti Empire** in what is today Ghana in West Africa. In the last war fought by the Ashanti against the British invaders, the defence of the Golden Stool was led by the formidable Queen Mother Nana Yaa Asantewaa, and gold is still a big issue in Ghana. Among the top producers in the world, Ghana exports more gold per square kilometre than any other country.

This huge export must, as the chemical industry is not so strong, be balanced by a corresponding cyanide import, mostly in the form of sodium cyanide, NaCN. Looking at this process from a different point of view we can write a new chemical equation:

$$2CN^-(\text{import}) + Au(\text{from ore}) \rightarrow [Au(CN)_2]^-(\text{intermediate}) \rightarrow Au(\text{export}) + 2CN^-(\text{waste})^{\dagger\dagger}$$

It looks now as to make one atom of metallic gold you will need two molecules of cyanide. And here, oddly, we meet Fritz Haber again because to make one cyanide ion you need one ammonia molecule, NH_3, the product of the Haber–Bosch process that combines hydrogen gas and nitrogen gas to produce ammonia.

Unfortunately, for the environment and working conditions in the gold industry, this simple equation is an oversimplification.

The reaction between cyanide ions and gold in water is instead what we call an equilibrium reaction. In theory, this means that the energy of the reactants is very close to the energy of the products. In practice, we can say that the gold atom needs to be surrounded by a good number of cyanide ions before it reacts. It is some kind of molecular see-saw for which we need a lot of CN^- to outweigh the noble character (or inertness) of the metallic gold particles (Figure 5.2).

In other words, the very unfavourable oxidation of gold is outweighed by the very favourable complexation reaction with

---

**The current English Wikipedia entries for the *Ashanti empire*, the *Inca empire* and the *Aztec civilisation* have 40, 5 and 3 occurrences respectively of "gold". Admittedly not a very scientific approach, but it might mean something. At its peak, the Ashanti empire encompassed an area roughly equal to that of the UK or the US state of Oregon.
††Showing only the most important components.

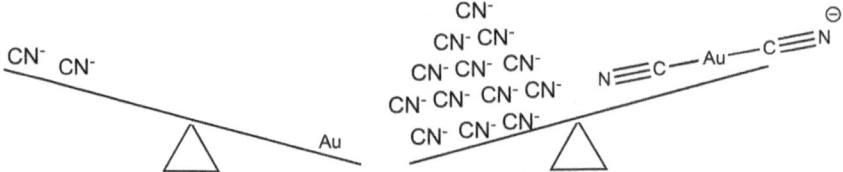

Figure 5.2   The gold and cyanide see-saw. To oxidise the gold particles and make the [Au(CN)$_2$]$^-$ needed to purify the gold, we must assemble many cyanide ions around the gold atoms, not just the two suggested by the formula.

cyanide ions. The equilibrium between gold, oxygen and Au$^+$ ions is so far displaced towards gold metal that only a miniscule amount of Au$^+$ ions are ever formed. But if these are continuously drained off by being transformed into [Au(CN)$_2$]$^-$ eventually all the metal will be dissolved, almost like a big dam being connected to a tiny pond. As long as the pond is not leaky, the level of the dam will stay the same, but as soon as the pond is connected to another big and empty reservoir, the first dam will soon be emptied. This will happen faster if we dig the connecting channels wider and deeper, which has its molecular counterpart in making the reactions move faster by increasing the cyanide ion concentration.

All in all, this means that the around 80 tons of gold exported from Ghana every year does not require 40 tons of NaCN, but instead around 27 000 tons, the amount imported in 2008.[1] Why is this unfortunate? Because NaCN is a very strong poison and it cannot, in a reasonable way, be recovered and reused once the metallic gold has been obtained. The whole handling of the NaCN, from import to waste is so problematic that the International Cyanide Management Code has been developed to make it easier for companies to follow best practice and handle this chemical safely throughout its lifecycle.

This is only one of the problematic aspects of gold mining in Ghana, both industrial and artisanal (the so-called galamseyers), but of course the industry also brings in revenues, both for individuals and the state.

The reason the cyanide ion is so poisonous is that it reacts rapidly and completely, no equilibrium here, with metal ions contained in important enzymes in our body. It does not, as carbon monoxide does, go for the iron ions in haemoglobin blocking the oxygen transport, but it blocks the ability of the *cytochrome C oxidase* to produce water (H_2O) from di-oxygen (O_2). This reaction, called a reduction, involves four electrons shuttling from the reduced *cytochrome C* to O_2 thanks to several metal ions purposely distributed in the protein core to bridge the outside with the inside of the mitochondria (the energy factory in our cells). In the conglomerate of proteins called *cytochrome C oxidase*, electrons are taken in one by one, first by a Cu atom directly bound by the protein, then by two successive iron atoms trapped in the heme ligands, as in the haemoglobin molecule shown in Figure 1.6, and ends up in the second Cu atom where O_2 is reduced. It is this consumption of oxygen that is called cellular respiration and more than 90% of the oxygen we breathe is consumed by this reaction.

During the travel of these four electrons, four protons are pumped out of the mitochondria. These ones and others can re-enter the mitochondria by a clever door able to synthesize the energy-rich molecule ATP. We understand now that the cyanide ion attack on these fundamental enzymatic systems causes a rapid failure in the energy production required for all our activities, without which the body rapidly closes down. The mechanism is still not completely understood, but it is likely that cyanide ions block iron in the +3 form in the *cytochrome C oxidase*, preventing the electron transport and thus cellular respiration (Figure 5.3).

Among the treatments and the antidotes used for cyanide poisoning, we find cobalt salts, used due to their ability to bind cyanide ions, as in vitamin B_{12} (more on this molecule later). However, cobalt complexes are also highly toxic and the advantages of such a treatment are questionable.

5.3   A GERMAN PARENTHESIS

Adding a H^+ ion to a negative cyanide ion will give the gas hydrogen cyanide, HCN. Hydrogen cyanide is also famous for its use by Nazi Germany to murder millions of people in gas chambers during the Holocaust in World War II. Here again is a connection to Fritz Haber, as Zyklon B, the nom de guerre of gaseous hydrogen cyanide, was developed by a former Haber collaborator, Walter Heerdt. By then Haber, himself a Jew, had

Figure 5.3   A cyanide ion (marked) binding to an iron ion in a model of *cyto-chrome C oxidase* blocking access and thus shutting down the generation of ATP from oxygen and sugar. ATP is the body's universal energy currency making a cascade of other enzymes work, all of which will now shut down and death could follow swiftly.

already died in exile in Switzerland. Designed as a pesticide, Zyklon B has never been used as a combat gas because of its low density, which causes it to spread randomly in the slightest wind. Of course, this was not the case in the confined space of the gas chambers.

Another name for hydrogen cyanide is prussic acid, but Prussia is also related to more peaceful compounds. The strong bonds cyanide ions form with some metals, for example to iron in *cytochrome C oxidase*, also work the other way. Adding iron to a cyanide solution will effectively scoop up and disarm the cyanides, transforming them into harmless iron cyanides, some of which we find in minute amounts in table salt as a way of preventing the salt crystals from growing too large (E535 and E536). We also find these iron cyanides in the pigment Prussian blue (also somewhat confusingly known as Paris blue), one of the first synthetic pigments with an interesting story of its own.

Although not a chelating agent, the structure of Prussian blue makes it very good at adsorbing cations with a charge of plus one, and it is in particular used to get rid of radioactive caesium ions (Cs^+) in the body. A major incident for which it was employed was the Brazilian 1987 Goiânia accident when a medical device containing radioactive caesium went missing and subsequently poisoned a large group of people.

Perhaps it would also have worked against silver 1+ ions too, as it works quite well as an antidote for the notorious thallium (Tl) +1, but the unfortunate Australian chemist, who accidentally had a silver solution sprayed in his face, did not have any Prussian blue at hand. But he guessed that if he let it be and just washed with water, the silver would still penetrate and with time give very distinct blue-black spots all over his face. So, he washed himself with a dilute cyanide solution that he had at hand. Not advisable perhaps, and certainly not under our current safety instructions, but 40 years later he was still alive and could tell the tale.

5.4 METALLOPHILLIACS OR MICROBIAL ALCHEMY?

Cyanides are of course, like so many other toxic chemicals, completely natural. More than 2000 different plants produce what are known as "cyanogenic glycosides", compounds that contain a sugar part attached to another molecule with a cyanide near the attachment point (Figure 5.4). When these come out of their biological confinement they may release hydrogen cyanide or cyanide ions when the different parts of the molecule divide up by reaction with water (what we call hydrolysis). Cassava is perhaps the prime example, and pre-treatment is needed to avoid the risk of poisoning, but also the seeds of apricots and flax may contain dangerous amounts of cyanogenic glycosides.[‡‡]

Plants themselves use this mechanism for a variety of reasons, one of course being as a pesticide, releasing HCN when attacked by insects. Plants are keen to produce a variety of different insecticides as they have difficulties outrunning their attackers. One other example is the notorious nicotine molecule.

[‡‡]So should not be chewed. But whole flax seeds will pass unharmed through the body without releasing any cyanides so these are not problematic. https://www.livsmedelsverket.se/livsmedel-och-innehall/oonskade-amnen/vaxtgifter/cyanogena-glykosider-och-vate-cyanid. One of the cyanogenic glycosides is amygdalin, found in apricot kernels, and a case of such cyanide poisoning was recently reported from Canada by CBC, Canada's public radio and television: http://www.cbc.ca/radio/asithappens/as-it-happens-friday-edition-1.4417898/man-treated-for-cyanide-poisoning-from-apricot-kernels-says-selling-them-like-nuts-is-nuts-1.4417904. See also the Australian guidelines: http://www.foodstandards.gov.au/consumer/safety/Pages/Apricot-kernels-raw.aspx.

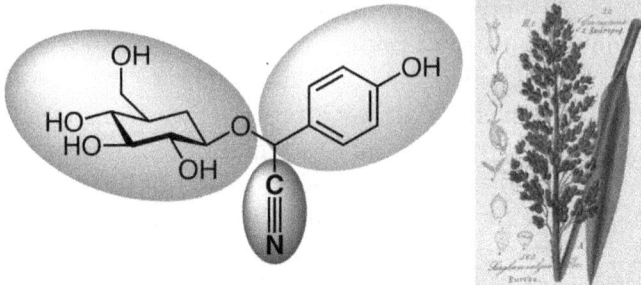

Figure 5.4   Dhurrin, from the Arabic for sorghum, was the first cyanogenic
glycoside discovered in Egyptian sorghum plants in 1902. After
this discovery, it could be proven that numerous cases of cattle
poisoning were due to the plant itself, and the release of cyanides,
and not an external fungus infection. Egyptians, of course, knew
this already. Dhurrin, like all cyanogenic glycosides, contain a
sugar part (left) attached to another organic molecule (right) with
a cyanide near the attachment point, easily released when the
molecule escapes its biochemical environment.

But humans mimic nature in more ways than one. Not only
are pesticides used by nature, but gold binding by cyanide ions
had already been practised in the biological world before it was
invented by a Scottish trio in the late 1880's. So instead of letting
the cyanide ions attack an intruding insect, they take care of an
intruding metal ion and transport it out of the organism.

This internally produced "chelation therapy" also uses bio-gen-
erated thiosulfate[§§] ions, the same ions used in classic photographic processes to remove unreacted silver from the developed
film or copy. In both these cases the CN^- ion binds the metal with
one atom only, the carbon giving the Au–CN units, and the thiosulfate ion often with one sulfur only, giving Au–SSO_3.

Could Fritz Haber's dream then be realised with microbes
digesting gold from seawater and producing tiny nuggets of solid
gold in some kind of microbial alchemy? This is still not clear,
but far from being an unreactive metal with no biological role,
we have now discovered that biology plays a major role in gold
geochemistry as these compounds play a significant role in moving gold around in the environment.

[§§]"Thio", deriving from the Greek for sulfur, is a somewhat archaic but still widely used
naming convention meaning that an oxygen atom has been replaced by a sulfur atom.
Thus, the sulfate ion is SO_4^{2-} and the thiosulfate ion is $S_2O_3^{2-}$.

5.5   GOLD, CRYSTALS AND WHITE OR BLACK COCAINE

In the last half of 2017, with only a few weeks in-between, two yachts were seized by the French Navy off the coast of New Caledonia (*Nouvelle-Calédonie*), a French overseas territory in the South Pacific Ocean. Did the marines have sophisticated analytical equipment on board? We don't know, but somehow, they figured out that the white powder on board was cocaine, a total of 2000 kg! One way they may have done this, using only a common microscope as analytical tool, would have been to react it with gold. Or to be more precise, acidic solutions of gold 3+ ions with four bromide or four chloride ions around. These $[AuBr_4]^-$ and $[AuCl_4]^-$ ions will quickly react with the cocaine molecule which even more quickly picks up an H^+ ion from the acid and is now positively charged. The reaction will produce crystals of a salt formed of metal based anions and organic cations, giving a characteristic shape of a "four-pointed star with approximately 45° angles" as indicated for example in the 2009 *Standard Operating Procedures* instructions from the Texas Criminal Law Enforcement department (Figure 5.5).

Possibly, this international gang of smugglers may also have tried to hide the white powder by making "black cocaine" using

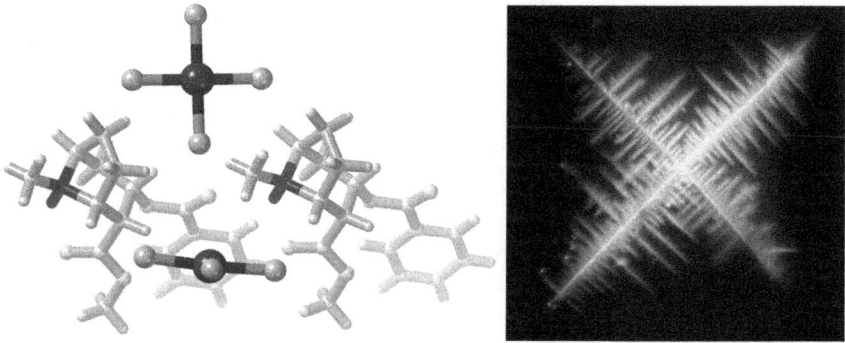

Figure 5.5   Characteristic crystals of the Hcocaine[$AuBr_4$] salt can be used for quick analysis of suspected substances. In the picture we show on the left the crystal structure of the square-planar $[AuBr_4]^-$ ions surrounded by positively charged Hcocaine$^+$ ions (with the nitrogen atom having the attached H^+ highlighted in black). To the right a microscopic image of the real crystals. Photo Dr Elizabeth Gardner, Department of Criminal Justice, University of Alabama at Birmingham, reproduced with permission.

a similar technique, but instead making a salt (as we call all compounds formed by combining positively charged ions with negative charged ions, for example table salt, NaCl) with the very similar copper 2+ ions surrounded by four thiocyanate ions,[¶¶] $[Cu(NCS)_4]^{2-}$. Just like the gold 3+ compounds it forms a square planar compound, but possibly, they could also have used the similar but tetrahedral nickel ion: $[Ni(NCS)_4]^{2-}$ for the same effect. With the risk, of course, that traces of nickel would set off allergic reactions in the customers. But perhaps that would not be a concern for drug peddlers.

The cocaine seized by *La Royale*, as the French navy is incongruently called, needed to be disposed of, and according to an article in *Le Figaro* the nickel industry of New Caledonia came to the rescue, and the powder was incinerated in a very, very, hot oven in a nickel smelter in the capital Nouméa.

5.6 STRANGE PLANTS IN NEW SCOTLAND

The choice of disposal method for the cocaine was perhaps obvious; the nickel industry is of huge importance for New Caledonia. With a total area of roughly 3% of the whole of France (or 0.2% of the USA), the country finds itself with 20–30% of the world's nickel reserve.[III] The mining industry is of course also an environmental concern, but the very nature of these islands contains a resource that can help. Two plants found only in New Caledonia, *Psychotria douareii* and *Sebertia acuminate* are known *hyper accumulators* as they can collect more than 3% of nickel (expressed as a percentage of dry matter) in the leaves of the *Psychotria* and the sap of *Sebertia acuminata* can contain up to 25% nickel, explaining the nickname blue sap for this species.

This is also known as *phytoremediation*, depending on if the focus is on getting rid of unwanted metal ions or collecting them for further processing. It is used both in field trials and in real life applications, from the environment around the nickel mines of New Caledonia in the Pacific to contaminated soils in Western

---

[¶¶]These ions form naturally in our body and the chemical procedure for getting the cocaine back is very simple.

[III]Counting ores that can be mined economically with the present technology and financial circumstances. This is different from the total resources on the planet. New Caledonia has around 9% of these.

Europe. Because it turns out this is not an odd phenomenon found only in the South Pacific.

The concept of phytoremediation was born in the 16th century and the first description was by the Italian physician, philosopher and botanist Andrea Cesalpino, of a plant called *Alyssum bertolonii* that was miraculously able to grow on the serpentine soils of his native Tuscany. Miraculously, because serpentine soils refer to soils with low silica content, a distinct lack of "NPK", that is nitrogen, phosphorus, and potassium,*** and instead high concentrations of what is called "heavy metals",††† especially nickel, iron, cobalt, and chromium. Phytoremediation needs hyper-accumulative plants, and among more than 400 known such species, more than 300 can harvest nickel at an average concentration exceeding 1% weight of the dried leaves and stems.

Therefore, paradoxically, the nickel polluted soils of New Caledonia support the development of a green chemistry sector involving revegetation with endemic plants species able to resist high concentrations nickel and other metals, to extract them and to store them in their green parts, mainly the leaves.

This works by plants sucking up various metal ions from the soil by first absorbing them into their roots. The metal can be taken-up alone, or already bound to natural or artificial chelators used to amend the soil and favour the soil-root transfer. Here again we meet the EDTA molecule used in chelation therapy in Chapter 3. The idea is that metal ions, for example lead, in otherwise insoluble combinations with other soil components, will be taken up by the EDTA molecule and then transported into the plant. The plant will be harvested and thus the metals removed. This so-called EDTA-assisted phytoremediation of lead-contaminated soil, or more generally, EDTA-enhanced phytoextraction of "heavy metals", is a matter of current research.

It is not clear if the metal-EDTA complex is then distributed in the sap moving water and nutrients from the roots to the leaves, or if the complex dissociates just after entering the plant

---

***From the Latin name *kalium*.

†††We put "heavy metals" inside quotation marks because it is a very problematic term and nobody knows what they are, at least 20 different definitions exist and the International Union of Pure and Applied Chemistry recommends not to use the term. Though we can be fairly sure that it is not H, Li, C, N, or O, beyond that it becomes tricky.

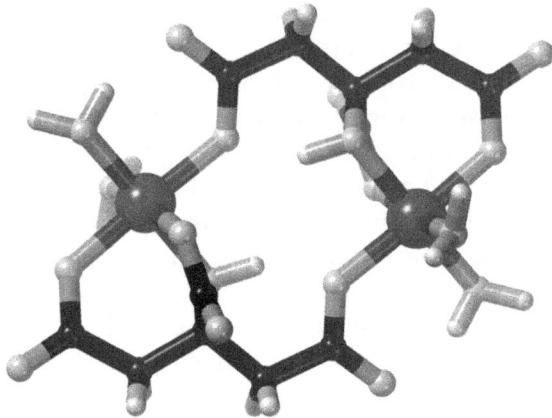

Figure 5.6 Citric acid, found in many plants, with lemon among them, can also help store nickel in the stems of *Leptoplax emarginata*. Two nickel ions with two anions of citric acid and some water molecules are shown.

root. In any case, once in the plant shoots, the metal is sequestered preferentially in the leaves or in the stems, depending on the species, and generally as a metal complex with organic acids. For instance, nickel is taken up by the hyperaccumulating plant *Peltaria emarginata* (a small flowering plant also known as *Leptoplax emarginata*) and ends up bound to the anion of citric acid (a citrate complex) in the stem and as a malic acid compound (malate complex) in the leaves (Figure 5.6).

Another type of molecule used to sequester toxic metal ions in plants are the phytochelatins. These are small molecules made up of a small number of amino acids, mostly the sulfur containing cysteine. They are not present in the plant under normal conditions but instead only synthesized in response to the increasing concentration of certain metals in the environment. This means there is a whole sequence of events taking place: the intruder ion must be recognized, its concentration determined, and the enzymatic machinery for the synthesis activated.

In the plants studied, cadmium seems to be the best inducer, far more potent than copper, zinc, mercury or lead. To be more precise, the phytochelatins are made from the tripeptide glutathione, in which the "thio" in the name means there is a sulfur[‡‡‡] atom present. This is the predominant thiol of living organisms, meaning it contains the –SH instead of –OH groups. It also contains the amino acids glutamic acid (the one in MSG, mono sodium glutamate) and glycine, the simplest and smallest of them all. Thanks to the thiol groups of the cysteines and of the carboxylate groups of the glutamic acid, they have a high affinity for metal ions and can play their protective role.

Coming back to phytoremediation, after phytoextraction, the cultivated hyperaccumulators are harvested, dried or incinerated before storage. Another strategy is to develop the valorisation of the phytoextracted metal, especially those with high added value. This strategy is called phytomining, the role of the plant being to concentrate the metal ions dispersed in the soil. As an example, an experiment in Ni phytomining has been conducted in the Balkans serpentine soils with the hyperaccumulator plant *Alyssum murale*. Up to 100 kg of Ni/ha have been collected and a potential of 200 kg ha^{-1} is expected by optimising all of the steps from culture to extraction. These processes can thus be both efficient and cost-effective, in addition to having environmentally friendly aspects.

5.7 A BEER CONCEPT TO HELP VINE GROWERS

We probably think of fertilisers mainly in terms of the NPK acronym, nitrogen (N), phosphorous (P) and potassium (K), but like humans plants also need "minerals" or "micronutrients", that are the metal ions that keep the biochemistry working.[2] Some soils are naturally rich in these, but sometimes they need to be added, and then we might get into trouble owing to the pH-value.

The cheapest way to added micronutrients is just to supply them to the plants as simple salts, iron sulfate, manganese sulfate or zinc sulfate, for example. These are very water soluble, so the plants should be able to suck the metal ions up from the soil efficiently, except some soils are formed partly by weathered

---

[‡‡‡]The Royal Society of Chemistry, the American Chemical Society and the International Union of Pure and Applied Chemistry all recommend the spelling "sulfur". The Oxford English Dictionary, however, insists on the spelling "sulphur".

particles from mountains such as the European Alps, or the White Cliffs of Dover, that is, they contain a lot of calcium carbonate. And calcium carbonate, a solid with an infinite array of Ca^{2+} ions with CO_3^{2-} in between, is basic. When dissolved in water it gives solutions with hydroxide ions, just like sodium carbonate (washing soda), but instead we have two Na^+ ions to balance the charge.

This is where pH comes in. You may have heard of soils being acidic or alkaline, and that this is measured by their pH-value. What we really measure is the concentration of H^+ ions, but to get from that to the pH-value takes some brewing.

A very strong acid like hydrochloric acid, HCl, will split up into Cl^- and H^+ ions in water, and therefore give very high concentrations of H^+ ions. A weak acid like acetic acid, CH_3COOH, in vinegar will only have a few of every hundred molecules split up, and therefore the concentration of H^+ ions will be much lower for the same amount of acid.

Pure water is the reference point and at 25 °C the H^+ concentration in that is 0.0000001 moles per litre. The unit chemists usually measure concentrations in is moles per litre, and 6×10^{23}[§§§] molecules make 1 mole, just as 12 eggs make a dozen. But to avoid making mistakes between 0.0000001 and, for example, 0.00000001 one must calculate the number of zeros very carefully. This was something Søren Sørensen at Carlsberg's brewery in Copenhagen got tired of very early in the 20th century. At the Carlsberg laboratory he worked with proteins and had found that these were very dependent on the concentration of H^+ ions, and he needed a convenient way to express this.

So, in 1909 Sørensen suggested that we write 0.0000001 as 10^{-7}, and that we then take the number 10 is raised to, -7 in this case, [¶¶¶] and multiply by -1 ($-1 \times -7 = 7$) and call this new number the pH value. Which means that the higher the H^+ concentration, the lower the pH value. For example, a 12% acetic acid solution that can be purchased in your supermarket has a pH of about 2, and pure water has a pH of 7.

[§§§]$6.02214076 \times 10^{23}$ elementary entities is the exact definition.

[¶¶¶]In technical language, we take the logarithm (log) with base ten. For example, $\log(0.001) = -3$.

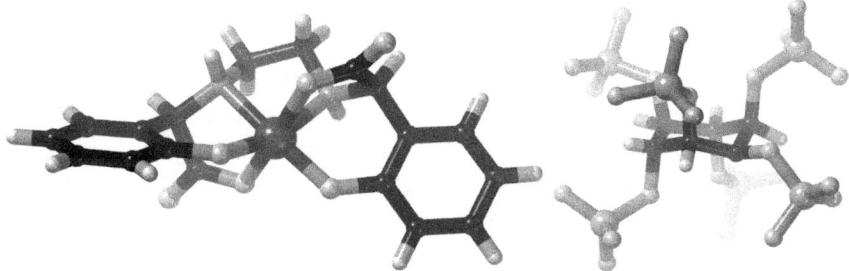

Figure 5.7 Micronutrient iron fertilisers, a so-called iron chelate. Left a common "synthetic" variant, Right: A natural chelating agent, phytate $C_6H_6(PO_4)_6^{6-}$.

Wine-growing often takes place in alkaline soils, and the plants need iron to survive. But the carbonate ions from the rocks will grab an H^+ ion from water, leaving a hydroxide ion behind:

$$CO_3^{2-} + H_2O \rightarrow HCO_3^- + HO^-$$

The carbonate ions will also chew up any free H^+ ions:

$$CO_3^{2-} + H^+ \rightarrow HCO_3^-$$

Thus H^+ disappear and OH^- appear. This is a problem if you like to have accessible metal ions in the soil. Iron ions will for example react with the OH^- ions forming insoluble iron oxides and rust particles. The solution is to fertilise, not with simple iron sulfates, but with an iron chelate that will stay soluble even at high pH (Figure 5.7).

In such a molecule the iron is well contained in an octahedron constructed by one organic molecule only. There is no escaping for the iron until it reaches safety inside the plant, where it can be release and the carrier molecules broken down. Needless to say, these products are more expensive than simple salts or "inorganic iron", and perhaps mostly used on plants that may yield products with a high price, such as grapes, because grapevines may easily become iron deficient in soils with higher pH values. However, they are also available for private consumers.

5.8 THE BIOCHEMICAL IRON-WAR

What could be the connection between the installation of biohazard detection systems in the US Postal Service major distribution centres at the beginning of the 2000s and the death of

1500 reindeers in Siberia in 2016? Objectively none, except for the causative agent.

On the 9th of October 2001, the senior biodefense researcher Bruce Ivins sent the same letter to two Democratic senators, Tom Daschle (South Dakota) and Patrick Leahy (Vermont). It seems that Ivins was looking for financial support to develop a vaccine against anthrax. It could have been a good initiative if the envelope had not contained a white powder identified later as *Bacillus anthracis* spores and had not caused the contamination of 22 persons, of whom 5 eventually died. This story followed a first set of letters sent to American media located in New York and in Florida and happened less than one month after the September 11 attacks. The anthrax letters referred rather explicitly to this horrific event. Evidence for the culpability of Ivins has been put into doubt. However, it is certain that the envelopes contained about one gram of pure spores of *B. anthracis*, the agent of anthrax infection.

The five people who died had all inhaled the spores and succumbed due to lung anthrax. This is the more dangerous form of the infection, the skin lesions caused by cutaneous (affecting the skin) anthrax are rarely lethal.

A spore (or to be more precise an endospore) is a sort of sinister bacterial version of the Sleeping Beauty. Some bacteria can transform themselves into this dormant and non-reproductive structure when the environmental conditions become inhospitable. Spores can survive without nutrients for extremely long periods of time and can revive when the conditions turn more favourable. If a spore enters an animal body, a human being for instance, through the skin, by digestion, or by inhalation suddenly it is like being kissed by the prince, and it will revive. The spore germinates and gives birth to a nascent bacillus that replicates, dramatically grows and eventually overwhelms the host, killing the prince.

The invasive stage of infection will continue with a phase during which the anthrax toxin will be produced, but this depends on the ability of the bacteria to find enough nutrients to multiply, and iron is a critical factor, in addition to a protein-rich environment. Iron is essential for most of the life processes, going from respiration to DNA synthesis, of a cell in general and of a bacterial pathogen in particular. However, the low iron availability makes a host an extremely hostile environment for the anthrax bacteria, which requires around 10^{11}–10^{12}-fold higher concentrations of free iron than is physiologically available in humans. The pathogen feels the host is an iron-depleted environment and the two organisms will engage in an iron war, the host maintaining severe limitations on free iron and the pathogens using specialized chemical systems for iron expropriation and acquisition.

Low iron concentrations will trigger the synthesis and deployment of specialized small organic molecules from the bacteria that have an extremely high affinity for Fe^{3+}, and therefore able to steal the metal from the host, even though it is hidden in a variety of complex ions. These specific iron robbing molecules are named siderophores (from iron [sideros] carrier [pherein] in Greek). Once loaded with iron, the Fe-siderophore complex is recognized by specific membrane transporters at the surface of the bacteria's cell wall and then uploaded into the cell's interior.

Within the cell, there are different mechanisms that make the metal available for the bacteria to use in its metabolism. One of the most effective, next to breaking down the ligand, is to transfer one electron to Fe^{3+} to convert it into Fe^{2+}. The direct consequence of this reduction is to make the metal less attractive (lower the affinity) for the siderophores so that they will spit them out and give them over to other organic ligands such as metallo-proteins or metallo-enzymes (proteins or enzyme whose function or activity depends on the presence of iron in active sites) for instance. Then the *apo*-siderophore (meaning the ligand without the metal) can be degraded or recycled.

In fact, most microorganisms, and also fungi and plants, use siderophores for iron-catching and feeding. However, in the particular case of bacterial pathogens, the strategies developed to access iron in deprived conditions should be considered as virulent traits. Incidentally, the development of new anti-virulent compounds targeting the bacterial iron metabolism to complete or replace the ageing antibiotics arsenal is a current matter of research.

The iron ligands of more than 500 siderophores already reported can be sorted into three main categories. One type of siderophore uses so called hydroxamate-groups, a bidentate ligand that can use two oxygens of the hydroxamate group to

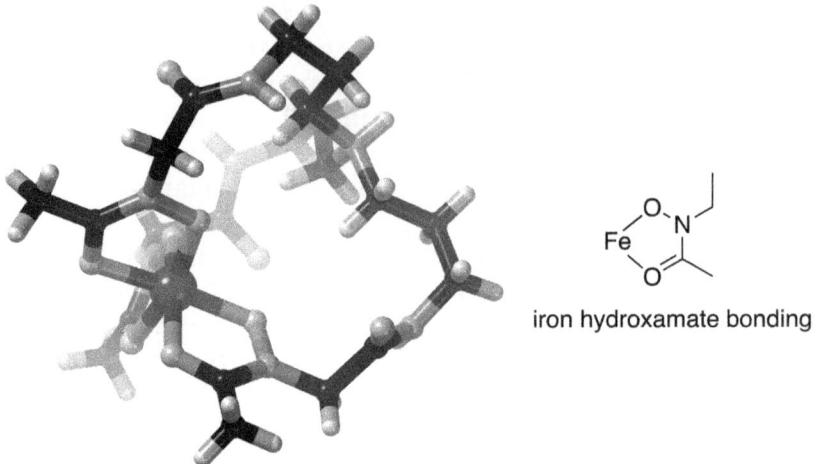

iron hydroxamate bonding

Figure 5.8 A metal extracting biomolecule, one of the siderophores of the hydroxamate type, specific for binding iron.

bind to iron, and three such groups together will form a very stable six-coordinated octahedral complex (Figure 5.8).

Siderophores using catechol, a chemical building block also known as benzene-1,2-diol, ubiquitous both in biology and synthetic chemistry, also make octahedral complexes, but the bonding atoms in this case are oxygens from the two neighbouring alcohol groups (1,2-diol), having lost their protons and become negatively charged catecholates (Figure 5.9). The archetypal siderophore in this type can be synthesized by the common bacteria *Escherichia coli* and is known as enterobactin or enterochelin.

The third category includes the carboxylate-type of siderophores that use carboxyl and hydroxyl groups to bind iron.

The anthrax bacteria synthesize two catecholate-type siderophores, bacillibactin and petrobactin, through two independent pathways.‖‖‖ In petrobactin the attachment point of the bulk of the siderophore to the six-membered catechol ring is to move just one carbon atom from the usual pattern in these siderophores.

‖‖‖The petrobactin siderophore is so named because it was first isolated from the oil-degrading marine bacterium *Marinobacter hydrocarbonoclasticus*. While the bacillibactin scaffold is based on classical 2,3-dihydroxybenzoic acid, petrobactin is constructed from the unusual 3,4-dihydroxybenzoic acid.

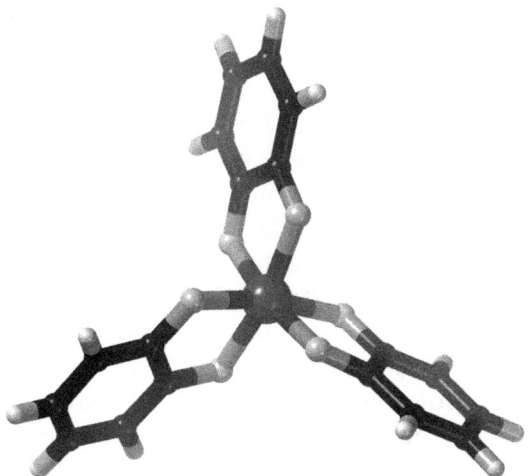

Figure 5.9 An iron 3+ complex with three catecholate ions $[Fe(catecholate)_3]^{3-}$. Catecholate based siderophores are responsible for how an anthrax bacteria steals iron from an infected human. Note the "propeller shape", we will get back to that in another context.

It does not make them better iron catchers. Both siderophores are 10-fold more efficient at chelating iron than transferrin, the major iron-binding protein of the blood plasma that regulates the level of free iron in our body fluids. Instead, this is a clever way to trick our immune defence. When infected, the host (us if we are unlucky) can defend itself by activating the immune system and producing a protein, siderocalin, that catches the invading siderophores and prevents them from stealing iron. However, because of its unusual chelating units, petrobactin escapes from siderocalin-binding and thus avoids the host immune countermeasure.

This means that the deadly force of the *B. anthracis* bacteria relies directly on petrobactin production and this explains why systemic anthrax (gastrointestinal or respiratory anthrax) is associated with high mortality. The current arsenal of drugs used to treat anthrax consist of combination therapy with two or more antibiotics. In the future, one could think of targeting the metabolic circuits of the petrobactin synthesis as an alternative anti-anthrax strategy.

In 2002, the US Postal Service installed biohazard detection systems at its major distribution centres to actively scan for anthrax being transported through the mail. In July 2008, Irvins committed suicide just before his indictment. Regarding the 1500 dead reindeers in Siberia, it appears that an anthrax outbreak was the cause. Anthrax was spread because of global warming that caused the region's permafrost to thaw with the consequence of melting the unearthed carcass of another reindeer that had died from a previous anthrax outbreak in 1968. Along with the deer, a 12 year old boy died and at least 40 people of the nomadic reindeer herders of northern Siberia were hospitalized.

More recently, more than a hundred hippos died in Bwabwata National Park in Namibia's Caprivi strip, bordering on Zambia and Botswana, probably because of a *B. anthracis* infection. Inactive spores of the bacteria can survive naturally in soil for years thanks to a hard coating helping them to resist heat, drought and radiation for long period of times. Spores can wake up after being ingested with grazing grass or by entering an animal through a cut or wound.

However, siderophores are not only the bad boys that connect iron metabolism with toxicity, virulence and lethal infections. They also play considerably peaceful roles in microbial ecology, in promoting plant growth, as biocontrol agents to regulate relationships between soil organisms, or to enhance bioremediation of heavy metals.

Last but not least, siderophores can have important applications in medicine. For instance, they have been disguised in Trojan horse type compounds to mislead the custom control machinery of bacterial cells walls, with the goal of delivering drugs into antibiotic-resistant bacteria. The strategy consists of the synthesis of antibiotic-siderophore conjugates with a spacer that will allow the antibiotic to be enzymatically released once inside the cell. In this way, chemists again try to imitate Nature.

For instance, some streptomycetes use albomycin in their struggle for an ecological niche against Gram-negative bacteria. Albomycin is a compound composed of an antibiotic moiety linked to a ferrichrome-like siderophore. The iron complex of albomycin crosses the membranes through the ferrichrome-uptake machinery before intracellular activation of the antibiotic moiety.

Another well-known medical use of siderophores is for the treatment of iron-overload diseases such as thalassemia, an inherited blood disorder revealed by abnormal haemoglobin production. To fight the resulting anaemia, patients are given regular blood transfusions that result in excess iron in body fluids. Iron overload may be treated by injection or oral ingestion of desferrioxamine derivatives, a hydroxamate-type siderophore synthesized by *Streptomyces* spp. Gram-positive filamentous bacteria found predominantly in soils and decaying vegetation. Incidentally, streptomycetes produce one-third of the clinically useful antibiotics of natural origin.

5.9 METAL-FISHING – A CONCEPT FOR A GREENER FUTURE

Synthetic and natural molecules designed to fish out metal ions from waste-water, leaching land-fills or mining waste with concentrations too low to currently exploit are going to be the major chemical innovations needed for the 21st century. We may look to the chemistry of the siderophores or other biologically inspired chemistry, but the solutions may well lie in completely unexpected directions as well. Chemical space, the different types of molecules we can make, how they interact and how we can use them, is still largely unexplored!

So, while Queen Mother Nana Yaa Asantewaa died in exile in the Seychelles 1921, the gold story is far from over. Maybe some cheap, naturally occurring molecules might be able to provide an alternative, much greener, process with the help of some very unexpected chemistry. As we shall see later.

Molecules that do not generate cyanide by the way. Many chemicals we extract from plants and trees are of course very useful indeed, and in the next chapter we will meet a few of these and see why we need elements from many parts of the periodic table to make them do their work.

REFERENCES

1. D. Engelbrecht, The Life Cycle Assessment of Cyanide Containers in Ghana, MSc thesis, University of South Africa, 2010.
2. H. Kiiski, H. Dittmar, M. Drach, R. Vosskamp, M. E. Trenkel, R. Gutser and G. Steffens, Fertilizers, 2. Types, in *Ullmann's Encyclopedia of Industrial Chemistry*, Wiley-VCH Verlag GmbH & Co. KGaA, 2016.

CHAPTER 6

Chemicals from Trees and How the Pacific Yew Was Saved

Historians argue about how many times Antonio López de Santa Anna was president of Mexico, anything from five to eleven apparently, but seem to agree that none of them were successful for anybody else but himself. In general, he is remembered for the Alamo, but even this 1836 military victory turned out to be a failure. To this we can add a third area where he did not do well, chemistry. The outcome was eventually a success, although not in the way he had planned and too late for him to profit.

6.1 GENERAL SANTA ANNA BITES OFF MORE THAN HE CAN CHEW

We humans are not ruminants, but still through the ages and cultures we have found different things to keep our jaws busy when we are not eating or talking. Natural polymers were our first choice for non-nutritional chewing, and in what is today Mexico, the Aztecs and the Mayans used *tsictle*, what we now call "chicle". Chicle is tree sap from local trees such as the *Manilkara chicle* and is tapped much like the sap from *Hevea brasiliensis*

The Rhubarb Connection and Other Revelations: The Everyday World of Metal Ions
By Lars Öhrström and Jacques Covès
© Lars Öhrström and Jacques Covès 2019
Published by the Royal Society of Chemistry, www.rsc.org

used to make natural rubber. It is in fact a lot like rubber sap, so when General Santa Anna had been ousted from Mexico for the umpteenth time and went into exile in New York in the 1870's, he brought a large supply with him with the idea of finding an ingenious US-American who could turn it into rubber. Hopefully making him rich again in the process so that he could bounce back to the Mexican presidency like a Mesoamerican rubber ball.

6.2   A FAILED REACTION

The process that Charles Goodyear had devised some 20 years earlier to make this amazing material that would vigorously spring back to its original form if you hit it, squeezed it or stretched it, did not work on the sap from the chicle trees, however. Not that Santa Anna dirtied his hands in the lab himself. It was the father and sons of the Adams family who tried to cross-link the linear polymer chains in the sap using sulfur, a process known as vulcanisation, making the material retain its original form instead of just having the chains sliding away in different directions like oily over-cooked spaghetti.

The circumstances of the meeting between Adams senior and the exiled general has been the subject of much myth and exaggeration, but it seems that Jennifer Mathews gets to the bottom of the story in her scholarly book *Chicle: The Chewing Gum of the Americas, From the Ancient Maya to William Wrigley*. Thomas Adams was, or became, a friend of Santa Anna's secretary Rudolph Napegy, who visited his glass store and learned of his work as an amateur inventor. Thus, the Adams family got their first samples of chicle from the General but could not get the Goodyear process to work.

There might have been many reasons for this, but one was surely the different chemical make-up of the saps. While the rubber tree has almost pure poly-(*cis*-1,4-isoprene) in its sap, the chicle tree yields mostly poly-(*trans*-1,4-isoprene) the difference being the same as between the much debated *cis* and *trans* fats, the orientation of the units attached to the carbon–carbon double bond (Figure 6.1). If the two biggest units are on the same side, then it is *cis*, otherwise it is *trans*.[†] The *trans* chains are

---

[†]It all happens because the double bond is stiff, you cannot rotate the *cis* form to the *trans* form without breaking the bond.

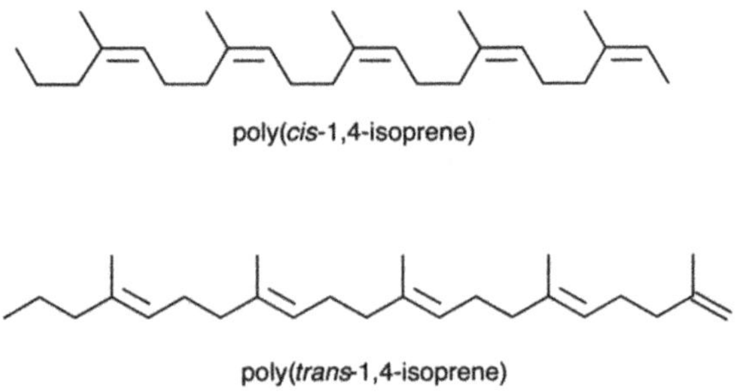

poly(*cis*-1,4-isoprene)

poly(*trans*-1,4-isoprene)

Figure 6.1 Basic structures of the two main forms of poly-isoprene. We have used the common chemical shorthand in which there is a carbon atom in each nick or branching point, and hydrogen atoms not shown. Top: Poly-(*cis*-1,4-isoprene) in the sap from the rubber tree, and bottom: poly-(*trans*-1,4-isoprene) from the chicle tree. The most common natural source of poly-(*trans*-1,4-isoprene) is the Malaysian Gutta-percha tree.

much easier to pack efficiently giving more regular and crystal-like structures, so our guess is that when the Adams family tried to vulcanize it they got something much closer to modern hard plastics than to rubber. Which at the time had no use.

To make a long story short, Santa Anna lost patience and eventually returned to Mexico, but never got close to any power again. Meanwhile the Adams family found they could make use of the chicle just as the Mayas did. They boiled it, moulded it to smaller pieces and sold it to drugstores, and when they started to incorporate sugar and flavours, sales really took off. By the late 1880's Santa Anna was long dead, but the Adams employed over 300 people in the world's largest chewing gum factory, close to the Brooklyn Bridge in New York.

6.3   INITIATING A NEW ERA

So far, the modern chewing gum was a success, but more was to come and that is where we will see the metals, in more ways than one. The Adams family were not the only ones making chewing gum, there was White, there was Beeman and, of course, there was William Wrigley Jr.

By the time World War II started three things coincided: the demand for chicle was now almost outpacing the growth of the trees, the advances in chemistry and chemical engineering made it possible to make the same, or very similar, polymers from petroleum products, and chewing gums became a standard item in the provisions for US soldiers.

Providing for the armed forces was such an undertaking that Wrigley's had to stop selling to the public. They were probably gambling on a world-wide demand to set in when the war was over, created by the US troops using their gums both for good-will and as means of payment in the war zones. This probably worked, as the current chairman, William Wrigley Jr II, has been among the top-600 on the Forbes list of the wealthiest persons in the world.

The high demand also meant a shift to a synthetic gum-base, the poly-(*trans*-1,4-isoprene) from the chicle sap, gradually, but not completely, being replaced with a synthetic variety. Today, there are a number of such materials approved by various food safety agencies around the world. The most common variety, then and now, seems still to be a polymer based on a mixture of butadiene and styrene. Butadiene is a molecule very close to the isoprene building block making up natural rubber, and styrene you might recognise, by the smell if nothing else, as one of the components of the ubiquitous all-repair agent known as plastic-padding (Figure 6.2). If that puts you off chewing-gums, you should know that styrene, as many industrially used chemicals, is also naturally occurring in cinnamon and coffee beans for example.

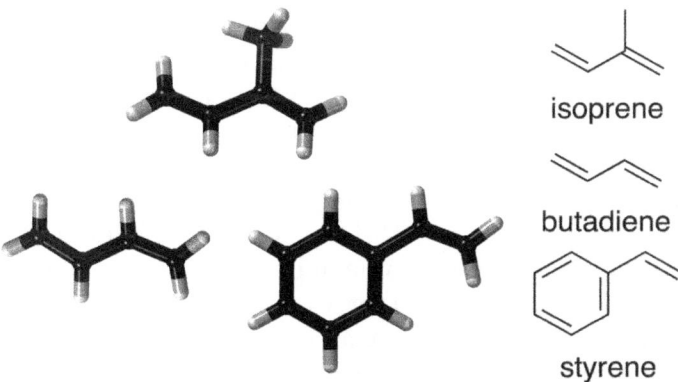

isoprene

butadiene

styrene

Figure 6.2   Isoprene, the molecular building block in both natural rubber and gutta-percha (see previous figure) and butadiene, the look-alike molecule that together with the last molecule, styrene, are the molecular building blocks of synthetic rubber.

6.4   SHADES OF BLACK AND GREEN

Was this move to a fossil based feedstock good or bad? This would take another book to figure out, but we would just like to note that very often the chemical side of these raw-material issues is forgotten. A few points can be made here: the current consumption of chewing-gum would not be sustainable if it were completely based on chicle trees. But then you could rightfully argue that the petrochemically derived gum-base is also not sustainable. On the other hand, if we were sensible and did not burn 90% of all the oil and gas we pump up from the ground, we would probably have another 1000 years or so of raw material supplies for all the chemicals we need. We would then have plenty of time to implement a bio-based economy, where we derive the chemicals we need from plants and other organisms. As it is, this development seems to be rather urgent.

This perhaps means back to the chicle tree. Not without complications either, however, land rights, conflicts over land-use, workers' rights, bio-piracy, all have been or are issues. We could try to dodge the question by saying that we do not need chewing-gum. Perhaps, but then again, this chewing has been going on in so many different cultures for probably 10 000 of years, that the masticating seems unstoppable.

6.5   METALS STRETCH THE RESOURCES

Be that as it may, synthetic rubber was developed neither to put Central Americans out of work, nor to make better chewing-gums. It was a 1930's German effort to become chemically independent of the British rubber tree plantations in Asia. Later on the Allies faced the same problem, as these plantations fell into the hands of the advancing Japanese army, and they used the same German process. With the additional benefit of providing cheap raw material for some of the 600 million chewing-gum sticks per year the US armed forces needed.

Making synthetic rubber was easier said than done, however, even though the structure and composition of the organic polymers had been worked out a decade earlier. (The idea that there existed molecules such as those in chicle sap with 10 000 repeating units was seen as quite outlandish early in the century.) The original process did not prove successful at first, as it was much too slow. Adding iron ions, however, proved to be the way to accelerate the reaction.

The process goes under the name emulsion polymerisation, implying that we do the reaction in some kind of dishwashing mixture. It is complicated, but the key is that the butadiene monomers, as we call the individual building blocks of the polymer, have two double bonds in both ends of the molecule. These can in principle be transformed into one central double bond and two odd electrons, one on each side. It will cost energy to create these lone electrons, but once they are formed they will rapidly react with any monomer and regain this energy and some. Formally we can see how this adds up by forming new bonds between the ends of different molecules, see Figure 6.3, but this is arithmetic only. In principle, we need to form only one radical to get the chain reaction going.

The speed of the reaction is dependent on how often two molecules bump into each other, and that, of course, depends on how many they are in the same space (we would say given volume). As only a few of the butadienes were initially transformed into radicals, the reaction was very slow. Adding iron 2+ ions with the chelating molecule ethylene-diamine-tetra-acetic acid (EDTA)

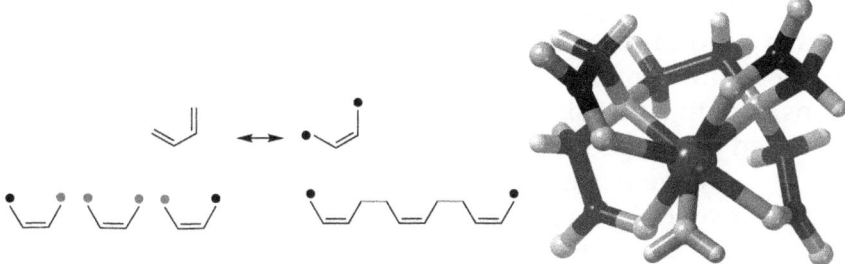

Figure 6.3 Butadiene, a cheap molecule from the petrochemical industry, but very much like the natural isoprene chemical that is the monomer in chicle sap and natural rubber, can have its two double bonds transformed into a radical, a molecule with one or more non-paired electrons (these are shown as dots above), and these radicals can combine with each other to form a polymer. The iron 2+ compound with the chelating ligand EDTA, helps to keep the free radical concentration high enough to keep the reaction speed up. This is a schematic picture, in real life we normally have only one odd electron per butadiene and the reaction proceeds stepwise.

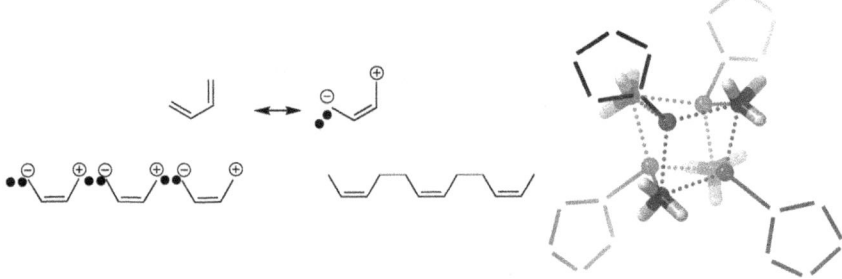

Figure 6.4 Butadiene is transformed into a molecule with one positive and one negative charge. Right is the methyl lithium molecule, [$Li_4(CH_3)_4$] the Li$\cdots$C bonds shown as dotted lines with organic solvent molecules are also bound to lithium (thinner lines). This is an extremely reactive organometallic reagent. The role of the lithium ions is to carry a negatively charged carbon in the methyl anion ($-CH_3$), this anion will immediately react with the positive end of the butadiene. The reaction proceeds in a step-wise fashion so that only one methyl anion is needed for a complete chain.

increased the free radical concentrations and at once made the reaction faster and commercially viable. The reason is that the iron can take up or give away single electrons, hopping between the oxidation states two and three.

Nowadays there is also another commercial way of making these types of polymers, anionic polymerisation. In this process, we can think of the two double bonds being transformed to one double bond just as before, but now the remaining electrons end up as a pair on one side, creating a negative charge, balanced by a positive charge on the other end (Figure 6.4). In reality, only the negative charge is created, as the positive side is taken care of by a rather exotic looking molecule containing four lithium ions. The role of the lithium ions is to carry a negatively charged carbon in the $-CH_3$ anion, which will immediately react with the positive end, and the reaction will proceed in a step-wise fashion.

6.6   TAKASAGO MAKES A MINT

The synthetic gum-base is slightly different from the "natural", but chemists also make chemicals that are completely identical to chemicals from nature. Just as a carbon dioxide molecule is

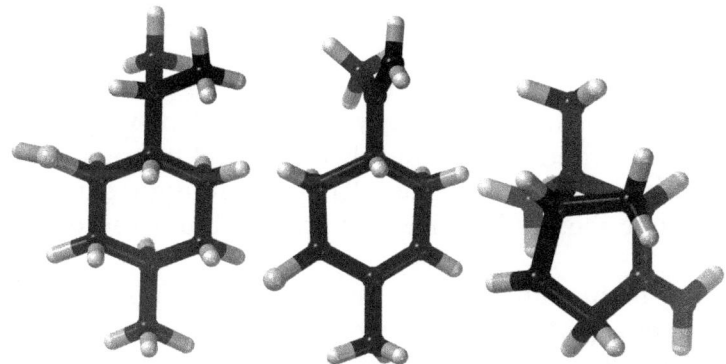

Figure 6.5   Menthol from peppermint (*Mentha × piperita*), carvone from spearmint (*Mentha spicata*) and beta-pinene from any kind of pine. Beta-pinene is a major component in turpentine and the source of many chemicals, including synthetic menthol.

the same whether it is exhaled with your breath, is created by yeast during bread-baking, is generated by baking powder in a sponge cake, or has been produced by burning lime to make cement.

One of these natural molecules is menthol, the major taste-creating molecule in Wrigley's doublemint chewing-gums. Doublemint, spearmint, and juicy fruit were the three Wrigley gum flavours the US soldiers in World War II got in their rations. The juicy fruit taste is somewhat undefined "fruity", the two others have clear tastes of peppermint, with menthol as the most important molecule, and spearmint, that contains carvone, a molecule with some resemblance to menthol and a similar bio-chemical origin (Figure 6.5).

The menthol in the WWII chewing gums most likely came from peppermint oil, but menthol is such a useful molecule that nowadays the demand seems to outstrip the natural resources. However, it is a bit trickier than just how much you can grow. Trade tariffs, taxes and other regulations make the balance between grown and synthesised menthol different in different parts of the world. Menthol finds its way into a variety of chewing-gums and other oral hygiene related products, in lip balm and cough medicines, as a topical analgesic in Tiger balm and other such products, in after-shaves[‡] and much more.

The current situation is that we now use thousands of metric tons of synthetic menthol every year,[§] so what are we going to do in the future when we cannot rely on oil and petrochemistry to turn out the molecules we need? The answer might surprise you, for synthetic menthol is already part of a bio-based or "green" economy.

The raw material needed for menthol is turpentine,[¶] which we get from the sap of various pine trees. This is a very abundant and renewable natural resource, and the more volatile parts, hydrocarbon molecules know as terpenes (made of isoprene units just like chicle sap) are released in the atmosphere from these forests in many thousands of tons every year.

Now it will get complicated, but just hold on and we will quickstep you through an organic synthesis scheme where we will go from one fragrant compound to the other. From the pine smelling turpentine, we get beta-pinene, which can be transformed into mycene by heating, which in turn in two steps (or should we say two separate cookings?) is changed into citronellal (a major component of citronella oil and an insect repellent), and then into isopulegol (a component of *Cymbopogon citratus*, commonly known as lemongrass) which finally gives us menthol. The Takasago International Corporation, which make 3000 tons of menthol this way every year, must have one fragrant-smelling chemical factory!

But where, you wonder, are the metals? Almost everywhere as it turns out. In the five final steps we need lithium in step one, rhodium in step two, zinc in step four and finally nickel in step

---

[‡]Menthol, for some time the only known cooling compound, was the starting point in what appears to be a continuing research project of the UK company Wilkinson Sword, a major manufacturer of razors and paraphernalia with roots in the 18th century. See J. Leffingwell and D. Rowsell, Wilkinson Sword Cooling Compounds: From the Beginning to Now, *Perfum. Flavor.*, March 2014, **39**.

[§]Judging from 2007 figures, the total world production of menthol is around 20 000 tons, 6000 of these being synthetic menthol. But then, of course, we have not included menthol used "as is" in peppermint and other sources, only the isolated chemical. G. S. Clark, *Perfum. Flavor.*, 2007, **32**, 38–47.

[¶]Note this is an unprecise term, there are many types of turpentines with varying origins and chemical compositions. See M. Gscheidmeier, and H. Fleig, Turpentines, in *Ullmann's Encyclopaedia of Chemical Technology*, Wiley-VCH, 2012, DOI: 10.1002/14356007.a27_267.

five. In all cases, except lithium, the metals are used as catalysts, so we need very little of them. A catalyst is not consumed in the reaction, it just helps along the union of the reacting species, a kind of molecular marriage-broker.

Which is lucky, because things do not get much more expensive than rhodium, which is used in the key step as we shall see. One time not long ago, you needed to dish out ten times as much for a gram of rhodium as for a gram of gold!||

6.7 IS LEFT RIGHT OR WRONG?

If you make a blueprint of any reasonable piece of mechanical machinery and take it to a well-equipped workshop with competent people, you can be fairly sure that you can collect the same item in real life within a practical time. Not so in chemistry. Menthol is a small, stable, and reasonable looking molecule obeying all rules of chemistry you may have learnt in school, but still it took quite an effort to devise a good way of producing it on a large scale other than extracting it from the peppermint plant.

One of the problems is its handedness. Our hands are peculiar, if you have not noticed. They look exactly alike, and yet they are not. Your left-hand glove will not fit on your right hand and *vice versa*, a mirror image of an object that is not identical to the object itself. That is, your left hand is a mirror image of your right hand, but as you discover when you try on the wrong glove, they are not the same. They are, as we call them, chiral.

This can happen to any kind of object, even very symmetrical ones. Propellers are typical examples, and in fact if they were not, boats would not move forward. Reversing what we call the chirality of the propeller, technical jargon for handedness and basically meaning the same thing but in Greek, would also make the boat go in reverse. Turn back a few pages to Figure 5.9 to see a typical propeller-shaped molecule, an iron 3+ complex with three catecholate ions $[Fe(catecholate)_3]^{3-}$, or to Figure 2.4 to see many "propellers" with alternating handedness linked together in the Siberian mineral stepanovite.

||The prices of course change every day and sometimes gold is more expensive, for up-to-date prices see for example: http://www.infomine.com/investment/metal-prices/.

This handedness thing happens a lot in the molecular world, and not only in life science chemistry. In beautiful *Rock crystals*, and other minerals based on quartz, silicon 4+ ions are bridged by oxygen 2− ions into an infinite three-dimensional network known as the quartz net. However, if you are lucky enough to find a nice piece of rock crystal while mountaineering, the chances are fifty–fifty that you will pick up a "left" or "right" handed stone, as these networks contain interconnected helices of the same handedness, and helices, just like propellers, are chiral objects.

Not so in biochemistry. All our amino acids have the same handedness and all DNA helices curve the same way. So, if our bodies are full of left-handed molecular gloves it obviously matters if an incoming molecule is left- or right-handed. For example, the limonene molecule found in its right-handed form in oranges smells strongly of (surprise) oranges, while the left-handed form found in mint oils has a piney smell.

6.8   A HELPING HAND FROM CATALYSIS

Menthol is also chiral, so one of the obstacles is to get the righthandedness in the final product.** This was for a long time a huge problem in the synthesis of organic compounds, which is one thing we have only mastered reasonably well in the last 30 years or so. The key to the Takasago synthesis is a rhodium compound of a kind that was a mystery to chemists for almost a century (Figure 6.6).

The compound contains one large organic molecule twisted like a two-bladed propeller, which binds rhodium through its two phosphorus atoms. This is all fine and not at all strange as the phosphorus atoms have a free electron pair each that can bind to rhodium, just as all other molecules we have seen so far. The odd part is the other side, where we have a pure hydrocarbon attached to rhodium. In this molecule, that will eventually become menthol, there are no free electron pairs, so how can

---

**It is even more complicated than left or right, as the menthol molecule has eight different possible forms (we call them stereo-isomers) and only one of these occurs in nature.

Figure 6.6 The catalyst is the key step in the Takasago menthol process. A rhodium 1+ ion binds to a large molecule which chemists for simplicity refer to as "binap" (top-side), which has two phosphorous atoms that each have a free electron pair that can attach to rhodium. On the lower-side there is a pure hydrocarbon anchored to the rhodium, where eventually the molecule that will become menthol will come in at the end of the game (no hydrogen atoms shown for this ligand).

it bind to the positively charged rhodium if it has no negatively charged side?

The origin of this catalyst lies in one of the great puzzles of inorganic chemistry, starting in 1830 with the Danish chemist William Zeise in Copenhagen cooking platinum chloride, $PtCl_4$ in ethanol (ethyl alcohol, or just "alcohol"), adding potassium chloride (KCl) and getting out a solid compound of formula $C_2H_6OKCl_3Pt$. The clever Zeise rearranged this formula to something like $KPtCl_3C_2H_4 + H_2O$, suggesting there was a molecule of water (H_2O) present (as there is always in ethanol,[††] we don't know what "proof" or concentration he used) and, a molecule of ethene or ethylene, C_2H_4, the smallest alkene.

[††]If you don't take to special tricks it is very, very, difficult to get rid of the all water in ethanol. The best one can do by normal distillation is 95.6% by weight, equivalent to one water molecule remaining for every ten ethanol molecules.

This was unorthodox bordering on the improbable, and some were shocked, including the great German chemist Justus von Liebig who in a vitriolic attack in a scientific paper called it a "fantasy".[‡‡1]

Nevertheless, Zeise was right but sadly did not live long enough to be able to rub Liebig's nose in it. It took some 30 years before somebody detected ethylene gas oozing off a heated sample of the material that is now known as Zeise's salt, but the mysterious missing free electron pair in the bond between the alkene and the metal was not truly understood until the 1950's. By then quantum mechanics had made clear that what we draw as two identical lines symbolising the carbon–carbon double bond C=C, is, confusingly, two very different bonds. One is formed by two electrons mostly buzzing around close to a straight line between the two carbons, where we perhaps intuitively would think they would be as that would seem to hold the atoms together by the positive nuclei on both sides being attracted to the negative electron "cloud" (the electrons do not really have specific positions at any given time) in between.

The second identical line we draw means something else entirely. The two electrons in this bond are spread out like two diffuse bananas, over and under the straight C–C bond. Not being trapped between two positive poles, they are more likely to wander further away, and if in doing so they find a positively charged metal ion close by they will attach to it.[§§] Not very strongly perhaps, but that is a good thing for catalysis, we want this molecule to be able to leave once it has metamorphosed to the right thing.

So, these goings-on are essentially that the rhodium compound, with the propeller like attachment with only one handedness, will catch a molecule of the starting material, fold it around itself, transform it to the right handedness, release it back into

---

[‡‡] A generation before, Goethe remarked something like: Chemists are nice people but just not to each other. Which seems to be a long tradition. In the 1660's Robert Boyle tried to persuade his contemporaries to be polite to one another in "The Sceptical Chymist" saying that: "...language should be more smooth and the expressions more civil than is usual in the more scholastic way". Apparently to no avail. But all our colleagues are nice people!

[§§] It is a bit more complicated as the metal also has a way of giving back electrons to the alkene and this was the key finding in understanding these types of compounds in the beginning of the 1950's.

the solution, and then revert to its previous form. It then grabs another one, and the process goes on and on, the very definition of catalysis being that the catalyst is not consumed in the reaction, it only facilitates it without being destroyed or changed.

So, what really is the best way of making menthol, through growing cornmint (having the highest yield of menthol) and extracting it from the plant, or from pine sap *via* a complicated chemical process? It seems a life cycle assessment favours the more chemical paths as they have less environmental impact.¶¶2

6.9   SAVING THE PACIFIC YEW TREE

In 1962, botanist Arthur Barclay, working with the US Department of Agriculture, collected samples from the Pacific yew tree (*Taxus brevifolia*) in Gifford Pinchot National Forest in Washington State. Tests of some extracts from the bark showed activity against cancer, and in 1979 Susan Horwitz could demonstrate that a compound in the extract, at the time know as taxol, had a hitherto unknown mechanism of action, a great asset for a potential drug molecule.‖‖ This was the starting point of a long journey towards clinical approval of the anti-cancer drug paclitaxel, still better known under the brand name taxol™. The trade mark, by somewhat mysterious ways, belongs to the pharmaceutical company Bristol–Myers Squibb. Paclitaxel, as we call this molecule from now on to avoid being sued, is very efficient in the treatment of breast, lung, and ovarian cancer, and on the World Health Organization's List of Essential Medicines.

If you thought menthol was a complicated molecule, have a look at paclitaxel. With over a hundred atoms it is no wonder perhaps that we had to wait a long time to find out how it looked! It took nine years on winding laboratory roads from the Washington woods to find the structure of the complicated molecule you see in Figure 6.7.

Is this another menthol case you wonder? Sort of, but much worse, or perhaps for a chemist better, as this is really a case of chemistry saving both the day and the Pacific yew.

¶¶The Takasago process is not the only process. There are at least two other ways to make menthol commercially from different starting materials, some from fossil sources.

‖‖Or API, active pharmaceutical ingredient. Not everything that goes into a drug is an active component.

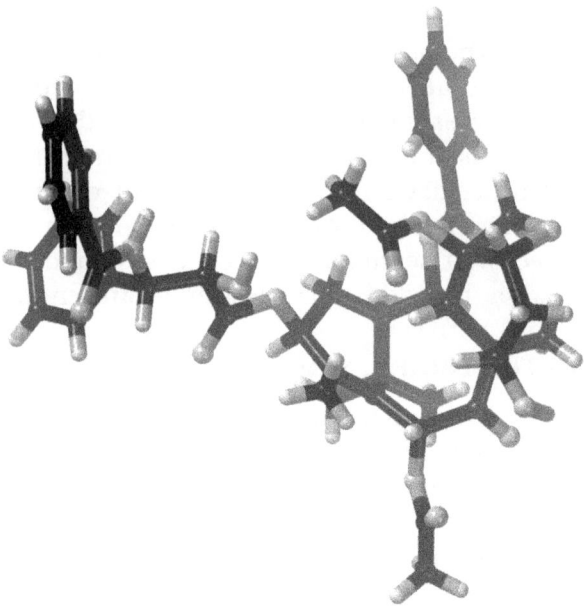

Figure 6.7 Paclitaxel is a large and complex molecule, very efficient in the treatment of breast, lung, and ovarian cancer. It is on the World Health Organization's List of Essential Medicines.

The problem was that there were not many Pacific yew trees. Also, they are slow growing and the concentration of paclitaxel in the bark is very low. In 1969, 1200 kg of bark finally gave medical research a mere 10 g of pure material to play with. Consequently, there was no way enough Pacific yew trees could be found or grown to sustain the demand that sky-rocketed when, in 1991, excellent results against breast cancer were reported.

What to do then? Just as for menthol, scientists looked at other plants, specifically other yew trees that might be easier to cultivate. They again did not find much paclitaxel, but in a close cousin, the much more common English yew, *Taxus baccata*, they found a molecule*** in a relatively abundant amount in the needles that could be converted to paclitaxel.

There are now various ways of producing paclitaxel, and more close cousins to paclitaxel have been found in yew trees. The needles regenerate and thus can be sustainably harvested for

---

***10-deacetylbaccatin III.

paclitaxel production. So, we do not need to worry about the Pacific yew becoming extinct anymore. And part of the reason is some clever chemistry involving metal ions.

One of the problems when operating on increasingly complex molecules is that making a desired change in one place might well change something for the worse somewhere else. It is like painting an increasingly complex plastic model of a ship or an automobile. To get the more exquisite details right, you need a very fine paint-brush in order not to mess up everything.

In chemistry, this translates into more and more selective reagents that will target specific parts of the molecule. That is why in one of these processes we meet a rather strange looking molecule, known as Schwartz's reagent, containing two zirconium ions (Figure 6.8).

The strangeness does not stop with the two zirconium ions though. These are actually not so strange, you can occasionally find them in your deodorant. The molecule also contains a metal–metal bond between the zirconium ions and has two negatively charged hydrogen ions bridging the two metals. We know

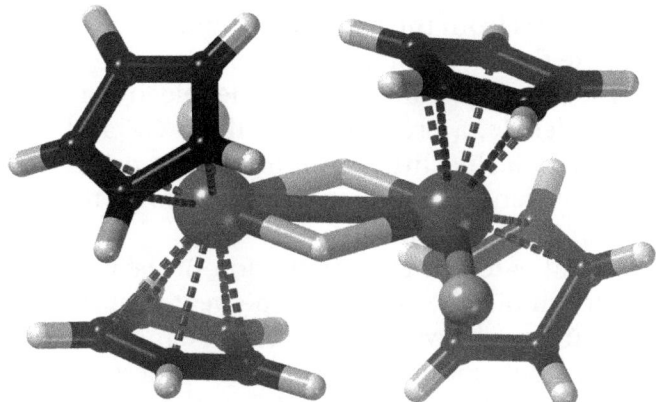

Figure 6.8 Schwartz's reagent, containing some unusual features we have not seen so far: it contains two metal ions, both zirconium +4, a metal–metal bond between these zirconium ions, negatively charged hydrogen ions bridging the two metals, and it has strange looking flat rings of five carbons and five hydrogens attached to the metals. This special bond is indicated by dashed lines. The only things that appear normal are the chloride ions binding to the zirconium.

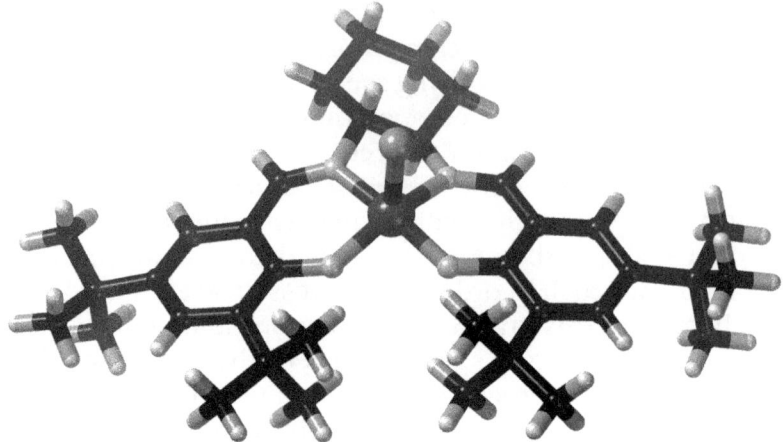

Figure 6.9   Deng and Jacobsen used this manganese catalyst to craft one
side-chain of the paclitaxel molecule. The single atom attached
to the manganese atom is a chloride ion.

these as "hydrides", and they are useful when we want to add
electrons to organic molecules, as the two electrons of the H⁻ ion
can easily attach to something else. Such as when making ethanol from acetic acid, CH_3COOH will become CH_3CH_2OH with
the extra electrons from the hydrides,[†††] and the hydrogen atoms
follow to make the new molecule complete.

   In that case though, we do not need the selective and expensive
zirconium molecule, a broad brush will be OK and it could be
some other hydride, like lithium aluminium hydride, the organic
chemist's equivalent of a reduction shot-gun, so reactive that it
catches fire in contact with air.

   An additional catch in the production of paclitaxel from the
closely related molecule found in the yew needles, is that an
amino acid fragment with a particular handedness must be
attached. This took some figuring out, but there are now many
ways to synthesize this amino acid fragment with the necessary
handedness and several of these methods make use of metal
catalysis. One such elaborate catalyst based on manganese is
shown in Figure 6.9.

[†††]The reverse reaction of this, obtaining acetic acid from the alcohol, occurs slowly if a
   bottle of wine is left open and oxygen (O_2), a powerful oxidant, finds its way in.

6.10 MEXICAN CHEMISTRY CHANGES THE WORLD

Arguably, the disasters of General Santa Anna changed the world. At the beginning of his rise to power, Mexico was a far larger state than the US. Afterwards, it had shrunk to its present-day borders, losing notably both California and Texas.

Another revolutionary change from South of the border was announced exactly 100 years after the Mexicans finally decided they had had enough of the incompetent general and threw him out in 1854.[‡‡‡] In 1954, an article was published in the Journal of the American Chemical Society, with the rather cumbersome and for the lay person indecipherable title "Steroids. LIV. Synthesis of 19-Nor-17alpha-ethynyltestosterone and 19-Nor-17alpha-methyltestosteron".[3] The second of these two molecules, better known as norethisterone, was first synthesised by Luis Miramontes, a 26 year-old Mexican chemist doing his undergraduate bachelor's thesis at the Syntex company in Mexico City. Norethisterone might mean nothing to you, but it was the first compound that made an oral contraceptive for women possible.[§§§]

Miramontes needed to do both reductions and oxidations to transform the naturally occurring estrone into norethisterone and, as chemistry was not so advanced back in the 1950's, he did not use molecular compounds such as we have described up to now. He used metallic lithium to do this reduction, the electrons coming from the lithium atom. But you need a very sophisticated trick to do this. First you need to make liquid ammonia. The ammonia we use for cleaning is a water solution but pure ammonia, NH_3, is a gas. So, it needs to be cooled down quite a bit to make it liquid. To -33 °C to be precise, then we add the metal.

[‡‡‡]Are we too harsh on Santa Anna? Perhaps, but checking with Mexican colleagues we got a clear message back, "no, no, that is fine, we all hate him!".

[§§§]The "first author name" on that article is Carl Djerassi, frequently considered to be the "father of the pill", but distributing scientific honours is a tricky game. Djerassi was a senior researcher at Syntex, but Luis Miramontes clearly made the compound first. Djerassi on the other hand gives a very interesting account of the whole story in his autobiography *The Pill, Pygmy Chimps, and Degas' Horse*, Basic Books, 1992. At the time of writing this book, the third chemist who co-authored that paper, former Syntex CEO and naturalised Mexican George Rosenkranz, was still alive, at 102 years old.

Figure 6.10   Very odd things happen when you dissolve a metal like lithium
or sodium in liquid ammonia (NH_3). You get a lithium 1+ ion
surrounded by four ammonia molecules, $[Li(NH_3)_4]^+$ and then
you get an electron all on its own! And this is how we know it
is blue.

Now you know what happens when you add sodium to water,
the favourite experiment of the primary school chemistry
teacher: The sodium floats and hovers on the water surface while
it gives away its electrons to H^+ ions coming from the water molecules. Two H^+ ions and two electrons will give one molecule of
hydrogen gas, H_2, the heat will occasionally be high enough for
the mixture of oxygen from the air and hydrogen from the reaction to ignite, giving off puffs of smoke and a bang. Lithium will
react in the same way, being sodium's upstairs neighbour in the
periodic table.

In liquid ammonia, there are no free H^+ ions to take up the
electrons given away by lithium, so what happens is that you
get a lithium ion surrounded by four ammonia molecules,
and an electron all on its own! We say that the electron is blue
as such solutions have an intense and beautiful blue colour
because of the free (or almost free) electrons in the solution
(Figure 6.10).

Adding an organic molecule to this solution will make the
electron go anywhere it might be welcome. In particular, the
electron will remove one of the double bonds in that extremely
stable building element called a benzene ring, and that is a really

tough job.[¶¶¶] Unfortunately, a double bond between carbon and oxygen (a ketone) in a completely different part of the molecule is also reduced so Miramontes needed to oxidize this to get back the essential ketone function having a C=O double bond. For this he used another crude shot-gun reagent, chromium trioxide, CrO_3.[∥∥∥]

Oxidation reagents do the opposite of reduction reagents, they take up electrons. As a rule of thumb, oxygen in compounds like CrO_3 has a charge of minus two and then, as the CrO_3 entity is neutral, the total charge from the oxygens equals minus six, which makes the charge on chromium plus six. This makes it suitable for taking up three electrons and ending up as the fairly innocuous Cr^{3+} ion.

If chromium six faintly rings a bell, it might be because of actress Julia Roberts. Chromium six, or as it was more often referred to in the movie *Erin Brockovich*, "hexavalent chromium", was the chemical culprit in this box-office hit that gave Roberts an Academy Award for Best Actress in 2001. The actual species in the Hinkley groundwater contamination, the subject of the movie, is the chromate ion, CrO_4^{2-}. This is a very efficient anti-corrosion additive and was used in a natural gas installation in the small town of Hinkley. It is the product you get after the negatively charged end (the free electron pair) of a water molecule has found its way to the very highly positive chromium centre in CrO_3, and then dropped off two H^+ ions to the water solution giving CrO_4^{2-}.

We have written about the Brockovich–Hinkley story elsewhere, so we will not make more noise about that here, except to note that searching for replacements for chromium-six reagents, especially catalytic systems that could use oxygen to take up electrons, is an active area of research in organic and inorganic chemistry. Neither will we comment on Roberts' obvious Mexico connection *via* the 2001 movie *The Mexican* and the former silver-mining town of Real de Catorce.

[¶¶¶]You may know benzene, C_6H_6, as a dangerous cancer causing chemical. True, but the building element with six carbons in a ring with alternating single and double bonds that we simply call a "benzene ring", we find everywhere in our own biochemistry.

[∥∥∥]This was, however, by no means the end of the chemical story of steroids. One could go on about the yam species *Dioscorea mexicana*, hearings in a U.S. Senate subcommittee, and extraordinary people like the African-American chemist Percy Julian.

Instead, we will return to Figure 6.8 and Schwartz's reagent, with the strange-looking flat rings made up of five carbons and five hydrogens attached to the zirconium ion. To chemists, such compounds were even more exotic than the funny platinum compound Zeise came up with in 1830. The first compound containing this strange unit, two hydrocarbon pentagons with an iron 2+ ion sandwiched in-between, is now known as ferrocene. We will taste the sweet success of this compound in the next chapter.

REFERENCES

1. H. Werner, *Landmarks in Organo-transition Metal Chemistry – A Personal View*, Springer-Verlag, New York, 2009.
2. D. Pybus and C. Sell, *The Chemistry of Fragrances: From Perfumer to Consumer: Edition 2*, Royal Society of Chemistry, 2006.
3. C. Djerassi, L. Miramontes, G. Rosenkranz and F. Sondheimer, *J. Am. Chem. Soc.*, 1954, **76**, 4092–4094.

Molecules and Murderers – The Art of Detection and Measuring

In the outskirts of Sydney, Italians Valter Molea, Riccardo Dei Rossi, Lorenzo Carboncini, and Carlo Mornati were furiously chasing four Britons on the Penrith lakes. Picking up speed at the halfway point, passing the US and Australian teams was, however, just not enough. The 2000 meter rowing competition known as the coxless fours[†] in the 2000 Olympics was won by the British team of Steve Redgrave, Matthew Pinsent, James Cracknell and Tim Foster.

This made Steve Redgrave a five-times Olympic gold medallist and earned him a knighthood from Queen Elizabeth II the next year. This was all the more remarkable as Sir Steve had been diagnosed with type 2 diabetes three years earlier. The key to the success was of course meticulous management of the disease, with controlled diet, carefully adjusted insulin injections, and measuring the blood sugar level up to ten times a day. All

---

[†]The coxless fours refer to a rowing boat with four rowers and no coxswain, *i.e.* that is there is not a special person steering the craft and overseeing the crew. Instead, a rudder cable is attached to the toe of one of the crew's shoes. The standard distance for all rowing races in the Olympics is 2000 m.

The Rhubarb Connection and Other Revelations: The Everyday World of Metal Ions
By Lars Öhrström and Jacques Covès
© Lars Öhrström and Jacques Covès 2019
Published by the Royal Society of Chemistry, www.rsc.org

of these, of course, being chemistry, but it is the last one that we shall discuss in this chapter. In order to do so, we will make a detour to Cambridge MA, the university town outside Boston that hosts Harvard University, and also go back in time some 60 years.

7.1   THE SWEET SUCCESS OF FERROCENE

It is unclear when ferrocene, $[Fe(C_5H_5)_2]$, was first made, but it seems to have been recorded as a 'yellow sludge' in the late 1940s by process technicians inspecting pipes at a Union Carbide cracker, which was used in the manufacturing of the small hydrocarbon cyclopentadiene from dicyclopentadiene. The technicians had no idea that this was something new and that it was going to revolutionise both organic and inorganic chemistry. They paid it no further attention but later it turned out to be the first example of a sandwich compound, a molecule having a metal ion squeezed in between two flat rings with five carbons having five hydrogens on the periphery (Figure 7.1).

Its first appearance in a scientific article a few years later caused quite a stir, as no stable molecule composed of only hydrocarbons and transition metals had ever been prepared. Even more astonishingly for chemists used to very reactive and pyrophoric organometallic reagents, it could be kept in air at room temperature

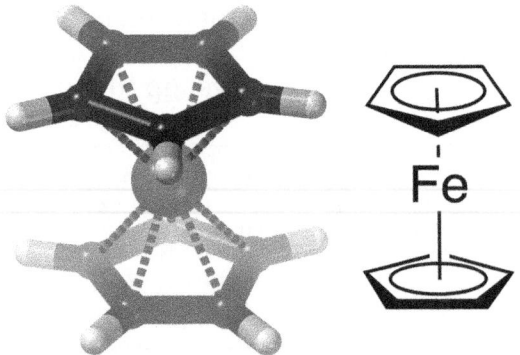

Figure 7.1   Ferrocene, $[Fe(C_5H_5)_2]$ as we now know it to look like. Left: a molecule from the crystal structure with dashed bonds between carbon and iron. Right: the way in which we normally draw this molecule.

without spontaneously igniting or doing some other nasty trick, something that made a large part of organometallic chemistry a very tricky business.

At this point, chemists knew the formula of ferrocene and some additional data, but had only vague experimental evidence of how the atoms in this molecule where held together. When the sandwich structure in Figure 7.2 was proposed by Wilkinson, Rosenblum, Whiting, and Woodward in 1952 it was like someone proposing the existence of zebras, giraffes and rhinos for the first time when you are used only to sheep, cows and horses. Although we have not found sarcastic comments in the scientific press, echoing Liebig's aggressive dismissal of Zeise's ethylene compound, the editor who handled this first article describing the "shocking" proposal for a new type of structure and new chemical bonds was a little taken aback. He wrote to the senior author Robert Burns Woodward saying: "We have dispatched your communication to the printers but I cannot help feeling that you have been at the hashish again, 'Remarkable' seems a pallid word with which to describe this substance".[1]

Figure 7.2   Chemical Landmark Plaque at the Inorganic Chemistry Laboratory at the University of Oxford. Photo credit: Karl Harrison Copyright 2018.

Ferrocene was the first example of the now very large class of compounds unceremoniously known as sandwich compounds (yes, we do call them that!) and more formally as metallocenes. They share not the triangular shape of the classic English sandwich, but the characteristic of squeezing something between two flat objects, namely a metal ion, the iron 2+ ion in the case of ferrocene. The two flat objects are cyclopentadienyl ions, each with a single negative charge. These are not triangular, but shaped like regular pentagons, making a molecule of unusual symmetric beauty. At least we chemists tend to think so.

Ferrocene owes its unusual stability to the fact that when one proton is removed from the cyclopentadiene molecule, C_5H_6, it becomes a flat and aromatic cyclopentadienyl ion, $C_5H_5^-$. Aromatic is a bit of a misnomer though, as this has nothing to do with aroma or smell anymore. Instead it means that apart from the ten electrons holding the ring together in five single carbon–carbon bonds, it has six extra electrons whooshing around above and below the pentagon. This is exactly the same situation as in the benzene molecule, the hexagonal C_6H_6, and the same explanation for its stability holds, partly due to the symmetric beauty we mentioned before. The similarity to benzene made chemists adopt the name ferrocene.

Less obvious is that sufferers of diabetes like Sir Steven often use this molecule several times a day. It is not a replacement for the small protein insulin, but instead a vital part of the electronic device that measures blood sugar levels from just a tiny little drop of blood with a minimum of fuss.

The device uses glucose oxidase, an enzyme used by some fungi and insects to produce the antibacterial molecule hydrogen peroxide from oxygen.[‡] This it does by catalysing the oxidation of a glucose alcohol group to a ketone, a C–OH becomes C=O just like when wine goes bad and sour. This adds two electrons to the enzyme and to complete the cycle it moves two electrons to an oxygen molecule (O_2), adds two H^+ ions lost by glucose to balance the charge, and voila, you get hydrogen peroxide, H_2O_2.

[‡]The human metabolism follows a different path with different enzymes, so that when we need hydrogen peroxide as a disinfectant, we need to produce it in chemical factories instead.

Back in the early 1980's a research group at the University of Oxford realised that if you glue the enzyme to a piece of carbon (graphite to be more precise) and take away the oxygen from the reaction, you will trap the enzyme in a state where it has two electrons too many. If you then attach a wire between the carbon chunk and the battery, you could in principle remove the electrons by electricity instead of oxygen. And when there is no more electricity going through the wire, then we must have used up all of the glucose molecules.

The good thing with this idea is that electricity is very easy to measure and can easily generate figures on a display. The bad thing is that it will not work, because we cannot transfer electrons directly to the electrode (this is what we can now call the piece of graphite carbon) from the enzyme, because these electrons are far inside the protein core. So, the enzyme will be stuck with its surplus of two electrons and the reaction stops.

This had been anticipated, and a solution contemplated. Perhaps a small molecule could be trapped in the graphite and enzyme matrix and used to shuttle the electrons between the electrode and the enzyme? For a lot of reasons, using oxygen and regenerating it by supplying the electrons to hydrogen peroxide will not work though. Both molecules are very reactive and oxygen is also not very soluble in water.

The device is instead loaded with ferrocene or ferrocene-like molecules.[§] By passing a current from the battery *via* the electrodes, these can easily be stripped of one electron to give a sandwich now containing Fe^{3+} instead of Fe^{2+}. Now the device is ready to measure, as these ions are small enough to reach deep inside the enzyme to pick up electrons from the reaction centre.

Drawing a tiny drip of blood and putting it on the disposable test-strip, every glucose molecule in the blood sample will now be oxidised, the enzyme will be temporarily reduced but very quickly[¶] regenerated by two Fe^{3+} in the ferrocenium ion and the battery supplies the electrons to get it back up to ferrocenium

---

[§]The first device used a ferrocene-type molecule. Today many different technologies exist on the market, as these are the most sold analytical instruments in the world. It has been argued that only pH measurements surpass blood-glucose analysis in the number of measurements made each day.

[¶]A very important issue is how fast these reactions occur, so it is essential to understand and mathematically model what we call the *kinetics* of the reactions.

again. By counting the number of electrons needed to re-oxidise ferrocene, that is the electrical current passed through the solution multiplied by the time it takes to use up all of the glucose molecules, the device then calculates how much glucose was in the blood and displays this number on a small screen.

And here we run into a completely unnecessary problem. The unit agreed by The International System of Units (SI) for concentration is kilograms per cubic meter, equal to grams per litre. Some instruments, often from the US, will instead use the non-SI unit mmol l^{-1}.[||] To mistake one for the other might mean the difference between life and death.

A final note on the voltage. Making the ferrocenium ion is not like rowing, but more like the high jump. If we want to measure how many good jumping athletes we have, we select a fairly high limit so that only the selected few will make it, and we add them up as they flop over the bar. This corresponds to a low voltage, only molecules that easily give up an electron will be pushed over the bar if the voltage is low. If the voltage is set too high other molecules can be oxidised too and then we are no longer measuring only the glucose concentration.

7.2 COUNTING ORANGES AND RAISING FISHES

Knowing what you measure is a central problem in the discipline of analytical chemistry. What is the selectivity of the chosen method of analysis and how can we be sure that we measure only what we mean to measure? To exemplify this with slightly bigger things than molecules, say we construct a device that will count the number of oranges in a bowl of fruit by passing them through a round hole, with each passing triggering a mechanical counter. This will work fine, as long as you have a mixture of say, oranges, bananas and pineapples. But if you also have apples in the bowl, chances are that these are now also going to be counted as oranges.

The solution might be a better counting device but we could also pre-treat the fruit bowl to get rid of the apples. For example,

---

[||]The very short list of non-SI countries includes the United States, Liberia and Myanmar. Moles per litre is fine when you need to make chemical calculations, but makes no sense whatsoever when communicating a concentration to the general public (1 mole is 6.022 $\times 10^{23}$ molecules and one mole of glucose molecules has the mass 180 grams).

we could give a first pass of the bowl of fruit to a peculiar variety of monkey that has an insatiable appetite for apples but hate oranges. Now all the apples, and some of the bananas no doubt, will have disappeared before we engage the counting device, and we will again get the right number of oranges.

One such procedure is used when we analyse the hardness of water, that is the sum of the concentrations of magnesium 2+ and calcium 2+ ions.** In everyday life, this is important to know so that you can add the right amount of powder when washing clothes, but might be more critical for big industrial installations. In hard water areas, we get a greyish scale of calcium carbonate forming in our electric kettle, a minor inconvenience, but the same phenomena can cause huge problems in industry, if precautions are not taken.

It seems that hard water was also an issue for young sockeye salmon that were raised in a hatchery in southeast Idaho and trucked to a nearby creek to be released. The appropriately named Clearwater Tribune reported in 2017 that the mystery of a low survival rate of the salmon might be due to an unexpected and more than ten-fold decrease in calcium concentration going from the hatchery to the release site. At least this is the current theory by the Idaho Fish and Game's scientist, having ruled out a number of other explanations.

Anyhow, determining water hardness is a standard water analysis method, normally reported when water quality analysis is performed. And although many analysis methods these days depend on fancy equipment and advanced instruments, obtaining the hardness is a classical type of analysis, something you might even have tried in school.

But first we need to do the monkey-fruit-sorting business, because the water will contain carbonate ions which will interfere with the analysis. These might either be from the rocks the water has passed on its way to the tap, or wherever the sample is taken, or just from carbon dioxide being captured from the air.

When carbon dioxide, CO_2, encounters water, an electron pair from H_2O will attach to the carbon atom and one of the C=O double bonds will break, forming a new molecule of carbonic acid. The carbonic acid will then part with an H^+ ion, making the water

---

**For simplicity, in reality water chemistry is more complicated than this.

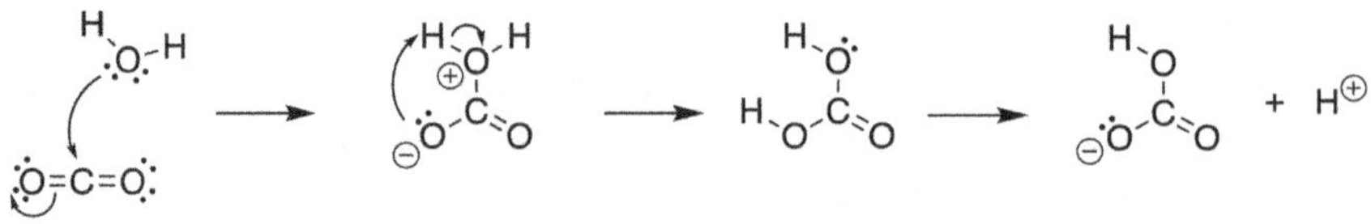

Figure 7.3   How carbon dioxide gas dissolved in water makes carbonic acid, and this then becomes a bicarbonate ion and H^+ ion, making the water it swims around in acidic. The curved arrows show the movement of electron pairs during the reaction, making and breaking bonds. In the first step, all valence electrons are drawn (as either dots or bonds), in subsequent steps only the electrons that move are drawn.

acidic and adding a bicarbonate ion, HCO_3^-, to the water (Figure 7.3). Incidentally, this is the chemistry that will make the oceans more acidic because of rising CO_2 levels in the atmosphere.

To get rid of the carbonate ions, we just do the reverse, we add an acid, transforming the carbonates to carbon dioxide again, and after also boiling the water, all of the gases will have disappeared, but the magnesium and calcium ions we want to analyse remain unchanged in the solution.

Now we need to "see" the Mg^{2+} and Ca^{2+} ions, and this we do by adding a ligand that will bind to both these ions, and produce a colour when doing so, an indicator molecule. This rather large organic molecule has the enigmatic name Eriochrome Black T™, and has, on its own, a dark blue colour when dissolved in water. If Mg^{2+} or Ca^{2+} ions are present they will be picked up by the Eriochrome and immediately the colour changes to bright red. Now we know that we have magnesium or calcium present, but we also need to know how much, and this is the next step.

Enter again the EDTA molecule, expert at enwrapping metal ions and making them invisible (first encountered in Chapter 3 and a drawing with calcium 2+ is shown in Figure 4.5). For simplicity, assume now that we have only ten Ca^{2+} ions and one Eriochrome molecule to start with. The solution will be red because the Eriochrome is bonded to one of the calcium ions. If we now add one molecule of EDTA, it will immediately react with one of the calcium ions. The solution still stays red, because there are still many other calcium ions that the Eriochrome can bind to. Adding new EDTA molecules one by one, the red colour will persist until the final Ca^{2+} ion has been engulfed by the 10th EDTA ligand. Now, there is no more calcium for the Eriochrome to bind to, and the colour changes to blue. As we have carefully counted the number of EDTA molecules we have added, we now know that there were exactly ten calcium ions in the sample from the beginning.[††]

Of course, we cannot count the EDTA molecules directly, but we can weigh them, calculate the number of molecules (or moles as this is our preferred unit) from that, dissolve them in water and then add small portions (that we keep track of) to the sample solution until it changes colour from red to blue. This is what we call a titration, and a general technique that can be used to measure a great many chemical things such as acids and bases, chloride, cyanides, ammonia, sulfide, and chemical oxygen demand, just to cite a few uses from the 2016 US Clean Water Act Methods Update Rule for the Analysis of Effluent.

7.3   QUANTITY OR QUALITY?

The hard water example embodies what might seem like a paradox. To ensure that we get a reliable result, we need to divide the reagent (titrator) into suitable small portions, smaller than the concentration we want to measure. That is, in order to measure the concentration, we need to first know the concentration! But this is no different from realising that we need different kinds of balances to weigh mice and elephants. One of the tricks in chemical analysis is to establish whether we are going to measure a mouse or an elephant to start with.

A way to tackle this problem is what we call qualitative analysis, we just establish that something is present, but not how much. Like finding elephant dung in your garden, you know then that there must have been at least one elephant present, but not how many exactly.

[††]In reality, this is a bit more complicated as we need to consider whether the reactions are equilibrium reactions and might not proceed to 100% products. A great deal of maths and statistics is needed to optimise such a procedure before it can be certified. For example, if we should accidently use an indicator that has a higher affinity for the calcium or magnesium ions than the EDTA molecule, then we will never reach the end-point, or reach it only with a massive surplus of EDTA giving a completely wrong value for the hardness.

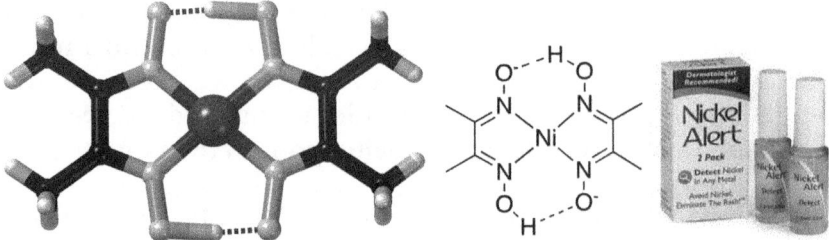

Figure 7.4   Two molecules of dimethylglyoxime, dmgH$_2$, having each lost one H$^+$ ion and now embracing a nickel 2+ ion. To the right is an example of a commercial dmgH$_2$ nickel spot-test kit sold to the public.

One such spot test, as we also call them, is used by botanic prospectors of potential hyper-accumulator plants. This test uses a compound able to chelate nickel ions with an immediate and dramatic colour change. This compound is dimethylglyoxime, dmgH$_2$ for short, as this molecule is an acid that can easily lose one or two H$^+$ ions. It is colourless, but turns pink to red when two molecules of dmgH$_2$ drop an H$^+$ ion each and then embrace one nickel 2+ ion, the colour is dependent on the concentration of the formed [Ni(dmgH)$_2$] molecules. Do-it-yourself spot tests with dimethylglyoxime are also sold so that you can yourself check for the presence of nickel in jewellery and other metals safely, quickly and easily (Figure 7.4, you might remember the nickel allergy discussion in Chapter 3.)

This test works on the same principles as the bullet-hole-test-kit described in Chapter 1: a metal ion is detected by adding a molecule (ligand) that produces a distinct colour change when the ligand engages with the metal. We will now return to the scene-of-crime for an example of the complete opposite, when a metal ion compound does the detection. Or not, as we shall see...

7.4   TO PLEAD OR NOT TO PLEAD, A QUESTION OF THE BLUES

We can call her Mrs A, a hard-working middle aged single mother from Louisiana. In 2010, her then boyfriend makes a dodgy manoeuvre with her car outside Houston, and a police cruiser signals them to stop. It turns out that her boyfriend does not have a licence, and from there on, her encounter with Houston's finest goes from bad to worse. Some white crumbs are spotted in the car, could it be drugs? The police officers perform a search, find nothing else of suspicion, but return to the police car to retrieve a small plastic bag with a vial containing a pink solution.

This is a spot test for cocaine, the pink solution getting its colour from cobalt 2+ ions surrounded by two thiocyanate ions, SCN^- (we saw them before in Chapter 5) and four water molecules to form an octahedral complex. One officer drops a few small crumbs into the vial, a blue colour appears and he waves it in front of Mrs A's face saying: "You're busted", because the blue colour indicates a positive test for cocaine in what is known as Scott's test.

There are a few variations of this test, some involving an organic solvent that will pick up the blue coloured component and leave a water phase that is still slightly pink, and it is used all around the world, being on The United Nations Office on Drugs and Crime's (UNODC) list of recommended methods. There are problems, however. The method is known to give "false positives" with a number of non-criminal substances, so a blue colour appearing in a field test would not hold in court as the only evidence. It is an indication only, and needs to be followed up with a proper identification of the substance by an accredited laboratory. If the defendant does not plead guilty of course, then immediate testing may seem superfluous and the evidence will just sit somewhere in a lab gathering dust.

Mrs A and boyfriend were promptly installed in different cells in the county jail, from where the young man was soon allowed to evaporate, never to be heard of again, as his only offence was driving without a licence. Mrs A, on the other hand, spent a sleepless night in jail, and when she saw her court-appointed defence attorney early the next day, he had an offer from the prosecutor: plead guilty now and serve some 20 days in the county jail, or face up to two years in prison.

Exactly what happened that morning is disputed, ProPublica reporters Ryan Gabrielson and Topher Sanders who wrote extensively on the subject in the New York Times Magazine in 2016, note that the defence attorney claims he also offered an alternative, where Mrs A could wait for further analysis of the alleged drugs before making up her mind, something Mrs A has no recollection of what so ever.

So, she pleads guilty not being able to properly consider the havoc even a mere 20 days in jail will have on someone already living on the margins. And she is not alone, in the US about 100 000 people plead guilty to drug possession based on field-test results every year according to Gabrielson and Sanders.

When we looked into the chemical side of this story, we were quite astonished to find that nobody really knew for sure what caused the blue colour to appear. One would think this should be a requirement for an approved test, as knowing how the cocaine molecule transforms to blue would greatly help in estimating what harmless substances would cause the same reaction, and thus be prepared against false positives.

We learn as undergraduates that the cobalt 2+ ion with six water molecules around it has a faint pink colour and that four chloride ions in a tetrahedral pattern around Co^{2+} gives an intense blue colour.[‡‡] Could something like this be going on? One way of getting a hint without going into the lab, and without going through the elaborate process of getting permission to work legally with cocaine, is to consider the 35 000 (and growing) structures of cobalt compounds in the Cambridge Crystallographic Database. One finds that for the large majority of Co^{2+} ions with two thiocyanate ions bound and having six atoms in total binding to cobalt, the colour is pink or red. On the other hand, when we have in total only four atoms binding to cobalt with two thiocyanate ions, we never get anything pink, but very often blue colours (Figure 7.5).

Checking out the cocaine molecule we see that there are two atoms, one nitrogen and one oxygen, that could conveniently form a chelate with cobalt, so we make an educated guess that the blue compound is $[Co(cocaine)(SCN)_2]$ which also should be nicely soluble in organic solvents, explaining the blue colour in that version of the test.

For this book, this guess is good enough, but there is really no substitute for going into the laboratory and exploring the real chemistry, preferably to obtain a good structure determination

---

[‡‡]This is the basis of the cobalt(II) chloride colour indicator in blue silica gel drying agents. When water is present we have the pink $[Co(H_2O)_6]^{2+}$ ion and when we remove all dampness, these water molecules evaporate and we are left with the intense blue $[CoCl_4]^{2-}$ ion or a mixture of different species (solid cobalt(II) chloride with no water is sky blue and is a network compound).

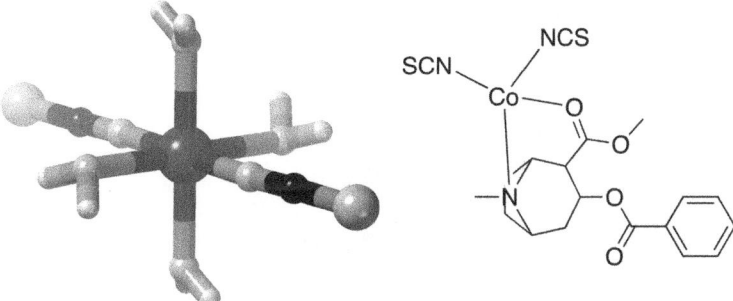

Figure 7.5   The pink [Co(H$_2$O)$_4$(SCN)$_2$] compound (this we know for sure, and the molecule is a real one taken from the Cambridge Crystallographic Database, as are so many others in this book) and an educated guess for the blue compound in Scott's cocaine test: [Co(cocaine)(SCN)$_2$].

of the compounds in question. Plus, growing crystals needed for crystal structure determination is a fun game, but please could someone else take care of the red tape?

So, no definitive answer there, but what about Mrs A? In 2014, the "drugs" had in fact been analysed and found to be NCS, "No Controlled Substance". It took a while for this result to get to the right persons, but in June 2016, the Texas Court of Criminal Appeals removed Mrs A's conviction, and a judge subsequently dismissed the charge a month later.

One of many such cases as it turns out. Having police officers in the field playing amateur chemists would perhaps not be so bad if the test was used to determine that whatever suspicious substance they find is *not* cocaine, and to get any blue-shining substances sent to a real laboratory pronto. "Busting" people based on this test only, and giving 100 000 people a year an "offer" to plead guilty before proper chemical analysis, is bound to land hundreds or even thousands of people in Mrs A's situation on a yearly basis.

7.5   THE CHEN–KAO-REACTION

A more well defined test for a regulated substance is the one devised by Chinese-American chemist and pharmacist Ko Kuei Chen (1898–1988) while at the University of Wisconsin in the 1920's. Chen pioneered the use of the ephedrine molecule in

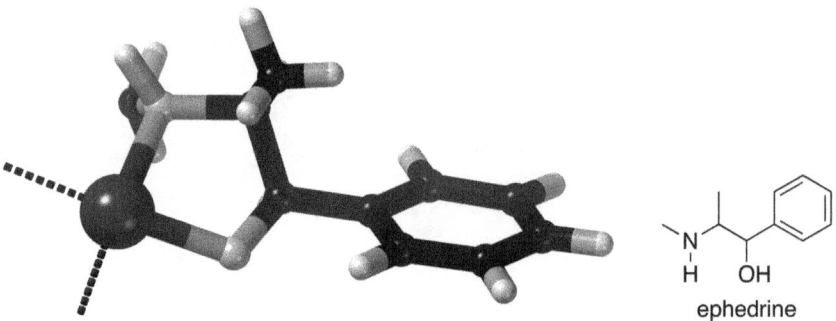

ephedrine

Figure 7.6 A Cu^{2+} ion binding one molecule of ephedrine, both a medication and stimulant, giving a characteristic violet colour used in the Chen–Kao test for ephedrine and related substances. The dotted lines show additional bonds to Cu(II), normally from another ephedrine.

western medicine, importing its use from traditional Chinese medicine. This molecule and its cousins are important as symptom relieving drugs against colds and allergies, for example. Working with inorganic chemist Chung-Hsi Kao (1901–1952), their Chen–Kao test simply employs copper sulfate, giving Cu^{2+} ions, dissolved in water and a base such as sodium hydroxide, NaOH. The latter will remove one H$^+$ ion from the C–OH group (the alcohol), making space for the copper ion to bind to two ephedrine molecules *via* both their oxygen atoms and nitrogen atoms (Figure 7.6). This is now a large neutral molecule and it is much more soluble in an organic solvent than in water. So, by adding an organic solvent that will not mix with water, one will see the violet colour move from the water to the organic phase.[§§]

Chen later became the Director of the Pharmacological Research at Eli Lilly and Company in Indiana and a professor at Indiana University, while Chung-Hsi Kao returned to his alma mater Tsinghua University in Beijing.

[§§]It also works for pseudoephedrine (violet colour), norephedrine and norpseudoephedrine (blue precipitates), as they all share the same OH–C–C–NH unit that makes the chelate bond with copper. These are also found in over-the-counter drugs, but pseudoephedrine is problematic, as it can easily be converted to amphetamine, an illegal substance.

7.6 A GLOW IN THE DARK

Science fiction writer Arthur C. Clark once remarked that "Any sufficiently advanced technology is indiscernible from magic", and the luminol test might certainly seem like magic. Get out a small spray bottle, and a few easy pushes on the handle will reveal a gruesome pattern of blood splatter on a seemingly clean and innocent surface in the form of an eerie blue glow, like something straight out of Harry Potter.

It works a lot better on television than in real life though, or at least it is visually more impressive than seeing it demonstrated in a real crime lab. But the information is there, no question, and you shouldn't really blame the movie people for making it more visible to us.¶¶

Can a spray bottle with a water solution be considered high-tech? Seeing how complicated this mechanism is, the answer is yes. No longer a question of molecule A meets molecule B and changes colour, this is a multi-step process with many components: the luminol molecule (Figure 7.7), the heme-molecule from haemoglobin in the red blood cells, hydrogen peroxide and a base.

The luminol molecule is colourless, but can be made to glow if it is oxidised. In doing so, it will pass through one of nature's oddities, an intermediate molecule that will break down and get rid of its excess energy, not only by rearranging its atoms and shaking off the energy as excess heat, but also by sending out a photon, a flash of light, as one of its electron drops from a higher energy level to a lower level.

We often like to picture reactions like getting something, say you and your bicycle, from a higher valley over a hill to a lower

---

¶¶Perhaps you can blame them for the way they dress some of the female CSI (Criminal Scene Investigation) officers but to insist that the science should be absolutely correct in all details, is in fact, preposterous. Often, producers do like to make things as correct as they can, but you cannot let scientific correctness derail the narrative. Science as an integral part of society demands that we scientists engage with the society around us also on their premises, not only on our own. That we might lack the willingness to do this can be exemplified by the lack of response the *Breaking Bad* producers faced when appealing to the chemistry community for expert help in getting the show's chemistry right. *Hollywood Chemistry – When Science Met Entertainment*, ed. D. J. Nelson, K. R. Grazier, J. Paglia and S. Perkowitz, ACS Symposium Series, 2013, vol. 1139. For *Breaking Bad*, see http://www.sonypictures.com/tv/breakingbad/.

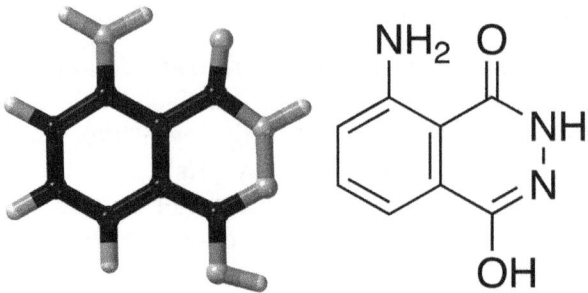

Figure 7.7   The luminol molecule from the crystal structure, and how we
normally draw it.

valley. If the hill is small, then this will be fast, if it is high, then
it will take longer. The height of the hill is a measure of how difficult it is for the atoms to rearrange to a form where a simple push
on the pedal will send you downhill to the other side.

Normally, all this potential energy that you have gained by
mounting the hill will now be transformed to heat when you
squeeze the brakes going downhill. Here, we are assuming there
is a conveniently placed pub in the next valley so that you actually want to come to a complete standstill at the bottom of the
hill. But, you could also attach a dynamo to one of your wheels
when you go downhill, and convert some of this energy to light
instead. As it turns out, very few molecules have dynamos, but
luminol is one of them.

However, we need an oxidising agent to oxidise luminol. This
is why hydrogen peroxide goes into the spray bottle. The reaction also works best if we remove the H^+ ions from the luminol
first, so the solution will also contain a base. Upon mixing these
components, there is nothing, no flash of light from the solution, because then it would be useless and used up far before we
get our CSI officer to the scene of the crime. We also need a catalyst, and this will be the iron-containing heme molecule from
the blood.

The catalyst will set off a rapid chain of reactions. First two
hydrogen peroxide molecules, H_2O_2, will be catalytically converted to oxygen, O_2, and two water molecules. We will now have
very high local concentrations of the powerful oxidant O_2 in the

thin layer covering the blood stain,[IIII] and this oxygen will react with the luminol molecules, breaking up the hexagon containing the two nitrogen atoms, blowing these off as nitrogen gas and in the process creating an intermediate molecule of the remains of the luminol that will be resting for a brief period on the top of a hill with the dynamo flat on the tyre. Then it will set off downhill, emitting a faint or bright blue light, depending a bit on the formulation and the circumstances.

Or something like it, for also this method can detect false positives. The difference to Scott's test is that the luminol method is used by properly trained scientific officers or personnel. Moreover, it is not used to coerce suspects into admitting guilt, it is more often used to find other more damming evidence, such as DNA traces taken from blood that we now know where to look for.

7.7   L'AFFAIRE FLACTIF

It is not difficult to find spectacular crimes where luminol was an important tool for the detectives in catching the criminals. Manufacturers of different blends of luminol also advertise their products by citing and describing successful cases. We will just briefly mention a case that is close to home, "L'affaire Flactif" as it is known in France, or the Flactif Case.***

In the small village of Le Grand-Bornand in the French Alps (2175 inhabitants), about halfway between Annecy and Chamonix, a mysterious disappearance occurred in April 2003. The brother in law of the real estate developer Xavier Flactif finds the Flactif's family home, an expensive *chalet* overlooking the

---

[IIII]Dioxygen, O_2, or oxygen for short, makes up 20% of our atmosphere and we cannot live without it, but it is also very insoluble in water. Luckily, as this is what makes water so good at extinguishing fires. Oxygen in its concentrated form is a very dangerous substance and only his knowledge of chemistry saves the protagonist in Isaac Asimov's *The Death Dealers* (1958, Avon, later republished as *A Whiff of Death*) from being blown up by a booby-trapped oxygen cylinder. The valve had been lubricated, which would have produced an instantaneous and fatal explosion, had he not noticed and refrained from turning on the tap.

***Written up in a book by French journalist Christine Kelly: *L'Affaire Flactif*, Calmann-Lévy, 2006. The 2012 movie *Possessions* by Éric Guirado is loosely based on the Flactif case.

ski-slopes, deserted when he arrives to spend a short holiday with Xavier, his wife, and three kids. He alerts the police who soon find the family car abandoned at the international airport in Geneva.

Mr Flactif has been of interest to the financial police for some time, he has an affluent lifestyle but his business is not doing so well. The first thought is that he has done a runner, together with the entire family, to a place where he has stashed away some money. But neighbours behave somewhat oddly, and it quickly becomes clear that Xavier Flactif and his business methods are despised by numerous villagers, one objection apparently being that he consistently outbids local buyers and then sells the property to Englishmen (or perhaps even worse, Parisians?) with a hefty profit. Xavier Flactif's origin in Guadeloupe, a French overseas department in the Caribbean, is seldom mentioned but may also have been a factor. Cautiously, however, no reference is made to the colour of his skin when some locals vent their frustrations with French press. These are on the scene *en masse* as a possible hideous crime committed in such beautiful surroundings is equally irresistible to both readers and journalists.

When the house is searched and screened by spraying with a luminol product, bloodstains are found in numerous places, although a thorough cleaning has obviously been attempted. When you know where to find the blood, DNA can be extracted and multiplied using the DNA polymerase method,[†††] and finally matched against known DNA from medical samples or known personal items. Shockingly, DNA from all the five members of the Flactif family are found in various bloodstains, leading to the conclusion that they have all been murdered in their own house.

In addition, mixed with the blood of the Flactif family is the DNA of a sixth, unknown, male person. Who could this be? The local police organise DNA testing of friends, family, work and business relations, and neighbours. In total 130 persons are

---

[†††]In 1993, the Nobel prize in chemistry was awarded to Kary B. Mullis for "the invention of the polymerase chain reaction (PCR) method." The enzyme in question, DNA polymerase, is dependent on an "active site", the place in the enzyme where the action takes place, containing two metal 2+ ions in close proximity. Normally these are magnesium 2+ (Mg^{2+}). More about the metals in our body chemistry are described in Chapter 9.

tested, around 10% of the village's male population. On the 8th of July there is a formal result, the unknown DNA belongs to one of the most outspoken and hostile neighbours, David Hotyat.

Cleverly, the police do not immediately spring into action. They are still piecing together a more complete picture using scientific evidence and can now also tap the telephones of Hotyat and his associates. In September David Hotyat is finally brought in, together with his partner Alexandra Lefèvre, and their friends Stéphane and Isabelle Haremza. None of them are native to the little village.

David Hotyat soon admits having committed the murders, his story, however, does not completely square up with the forensic evidence. He later retracts the story in court, now claiming two unknowns he is too scared to describe committed the murders. He admits to doing away with the bodies and the evidence, and the remains are eventually found and identified. The three others, however, admit to a conspiracy a long-time in the making, but not to murder, and how things exactly happened cannot be established by the forensic evidence alone. Still in prison for life, David Hotyat keeps mum, and we might never know the whole truth.

Stéphane Haremza was convicted to 15 years in prison, Alexandra Lefèbvre to ten and Isabelle Haremza to seven. David's brother Mickael Hotyat, who got rid of the gun that was used in the killing of both parents, got away with a one year suspended sentence.

7.8   A COLOURFUL DEATH

Some murders, real and fictional, are committed by poison, and the detection of these in the dead body made major advances in the 19th century. The Marsh test for arsenic was one, producing a nice silvery coating on the test tube used by Lord Peter Wimsey's manservant Bunter in *Strong Poison*. A much more colourful metal-oxide based test was developed for the natural organic product strychnine. Professor Raychelle Burks, who has been kind enough to write a foreword to this book, recently dug up the gloomy Christmas tale of Chesapeake Bay farmer William Taylor, and his wife Virginia and wrote about it in Chemistry World.[2] At the time, 1888, the most recent of these tests was Mandelin's, in which tetrahedral vanadate ions, vanadium 5+ ions surrounded by four oxygen 2− ions, gradually grab electrons from the strychnine molecule, moving from the 5+ ion to the more stable 2+ oxidation state while producing a characteristic sequence of beautiful colours, the exact composition and molecular structure of these different compounds again not known. Vanadium, by the way, was named after the Nordic goddess of beauty (and other things) Vanadis or Freya.

7.9   THE HOLE IN THE TABLE AND AN UNFORTUNATE CHAIN REACTION

Detecting chemical components from body fluids is an essential tool for CSI teams, now as in the 19th century, and detecting what goes on inside a body that does not feel completely well is another form of detective work, performed by physicians and hospital scientists of various specialities. Some of these were severely worried when the old nuclear reactors at the Chalk River Laboratories, Ontario, Canada and at the Energy Research Centre of the Netherlands in Petten[‡‡‡] were closed for maintenance repairs and safety for long periods starting in 2009.

But we should really start this story with a gigantic hole in the periodic table, and a remarkable female German chemist Dr Ida Tacke (1896–1978, married as Noddack). In 1925 the periodic table started to look complete, but there was a large hole among the elements known as the transition metals (see Figure 7.8), two elements, with atomic numbers 43 and 75, seemed to be missing under manganese. Ida Tacke and her husband-to-be Walter Noddack, together with co-worker Otto Berg, tried a real tour-de-force in their article Ekamanganeses (*Die Ekamangane*, "eka" being a now obsolete way of indicating an unknown element below a known element) that year. They presented two new elements at the same time, masurium with atomic numbers 43 and rhenium with atomic number 75.

Rhenium made it to main stream chemistry, Noddack, Tacke, and Berg are hailed as the discoverers, and you can now buy

---

[‡‡‡]Managed by the Nuclear Research and Consultancy Group part of the Energy Research Centre of the Netherlands (ECN).

Sc	Ti	V	Cr	Mn	Fe	Co	Ni	Cu
Y	Zr	Nb	Mo	?	Ru	Rh	Pd	Ag
Lu	Hf	Ta	W	?	Os	Ir	Pt	Au

Figure 7.8   The transition metals of the periodic table as it was in 1925. A gigantic hole below manganese (Mn) where there should be two elements with atomic numbers 43 and 75.

rhenium compounds from major chemical suppliers. Masurium, however, drifted into obscurity, and is now of interest only to Periodic Table aficionados. On the contrary, element 43 lives on, and the New York Times has estimated that 40 000 medical procedures a day use this element in the United States.

How does this square? The easy answer is that there was simply no element 43 in the samples and the faint signal Tacke and her co-workers detected must have been something else, or perhaps just wishful thinking. The more complicated answer is that there were loads of other reasons for the scientific community to reject the claim, including the political unsavvy choice of a name derived from Masuria, a former part of East Prussia (Germany), and now part of the free Poland that had re-emerged after World War I.

Unlike almost all other dubious claims for new elements presented during the late 19th and early 20th century, the claims of young Ida Tacke,[§§§] she was 29 at the time, and co-workers continue to be discussed. There have been arguments, based

---

[§§§]Ida Tacke should perhaps be even more famous for being the first to mention the idea of nuclear fission in a 1934 article with the modest title *On element 93 (Uber element 93)*, an understatement as subtle as they ever make them in Britain. She tears apart a paper by Enrico Fermi, where he makes an erroneous claim for element 93 (maintained by Fermi even in his Nobel lecture four years later: the real element 93, Neptunium, was to be discovered during WWII and the Manhattan project) and as a colleague of ours bluntly put it on twitter, "Fermi got the 1938 Nobel because he and his buddies didn't read chemistry papers". Something else that worked against Tacke was of course a somewhat tainted reputation after the discredited claim for element 43, although she was nominated three times for the Nobel Prize in Chemistry. An interesting discussion is found in Gildo Magalhaes Santos, A Tale of Oblivion: Ida Noddack and the 'Universal Abundance' of Matter, *Notes Rec.*, 2014, **68**, 373–389, (the Royal Society Journal of History of Science).

on what we know now, that perhaps they could have been right after all. However, the latest conclusion seems to be that this was indeed an erroneous claim.

The real element 43 was discovered some 10 years later and was, as it seemed it could only be produced in any reasonable quantity by using nuclear reactions, definitively named technetium in 1947, with the symbol Tc. More relevant to our story though, the isotope technetium-99m was made in Berkeley from molybdenum[¶¶¶] by Glenn Seaborg and Emilio Segrè in 1938, and it turned out to have near perfect properties for use in medical diagnostics.

Tc-99m emits X-rays of about the same wavelength as conventional X-ray generators in hospitals, easily detectable by a scintillation camera, also called a gamma camera. If we have done the right chemistry beforehand, the properties of the various Tc compounds will send them to different places in the body that we need to investigate. It also has a half-life, the time it takes for 50% of the technetium-99m nuclei to decay, of 6.6 hours which enables relatively quick measurements involving the diagnosis of thyroid, liver, brain and kidney disorders. For example, a MAG3 scan to investigate kidney function uses a technetium compound in which the ligand wraps around the metal ion, see Figure 7.9.

And what is now the connection to the ailing old nuclear reactors? It turns out that we need installations like that to produce enough technetium-99m to provide the world with the material for the 40 million or so medical procedures per year in which it is used.[‖‖‖] A worldwide supply crisis erupted around 2009–2010, that now seems to have been resolved.

[¶¶¶]Simplified you could say that all you need to do is to add a proton to the most common isotope of molybdenum, molybdenum-98 (having 42 protons and 56 neutrons) to get technetium-99m with 43 protons and 56 neutrons. The reality is more complicated, however. The designation "m" stands for meta-stable, there is also another technetium isotope with the same number of neutrons and protons that has a half-life of 211 000 years.

[‖‖‖]More than 10 000 hospitals around the world use radioactive compounds such as technetium-99m (the most common), mostly for diagnosis. http://www.world-nuclear.org/information-library/non-power-nuclear-applications/radioisotopes-research/radioisotopes-in-medicine.aspx, World Nuclear Association, seen 11 Dec 2017. Other isotopes routinely used for diagnosis are for example iodine-123 and 131, thallium-201, gallium-67, fluorine-18, and indium-111.

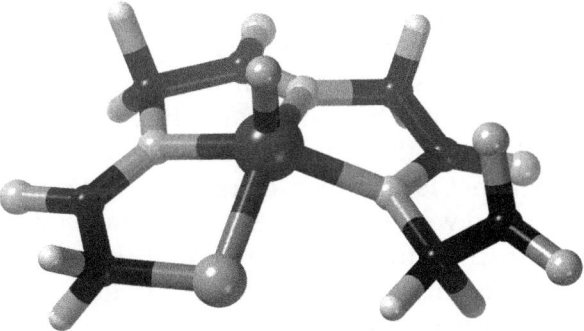

Figure 7.9 The mercapto acetyl triglycine (MAG3) wraps itself around the technetium 5+ ion having an additional oxygen doubly bound (we call that a technetyl ion, TcO^{3+}). This is a useful compound for inspecting your kidney function.

7.10 NOW YOU SEE ME – MUCH BETTER

Chuck Norris and Mark Ruffalo appear to be two very different Hollywood actors, despite a certain muscle emphasis, in the case of the latter being impermanent and greenish. One has republican leanings, the other democratic. One, at the time of writing, seems to be taking on "big pharma" in a legal battle, while the other would probably not be acting anymore without them.

Ruffalo woke up one morning after dreaming of having a brain tumour, something that unsettled him so much that he immediately set off to see his physician. Oddly, and sadly, enough he was right, a tumour as large as a golf ball had been growing close to the nerve that transmits hearing and balance information to the brain.**** The standard way of detecting such a tumour is gadolinium-enhanced nuclear magnetic resonance imaging, usually only referred to as MRI, as the little word "nuclear" might give people the wrong associations.

It is a wonderful way of peeping inside the body, but in contrast to the technetium methods, MRI has nothing to do with nuclear reactions or radioactivity. Instead, we have completely normal radio frequencies and strong magnetic fields. And we do not look at some exotic atom like technetium, but at the hydrogen atoms of water molecules.

****The correct medical term seems to be vestibular schwannoma (VS), also called an acoustic neuroma.

The hydrogen atoms, or to be more specific, the ones with one proton and zero neutrons in the nucleus, as have 99.99% of all hydrogen atoms, behave like small compass needles, but the term we use is that they have nuclear spins. Put them in a magnetic field and they will all point north (or whatever direction the external magnet is pointing) or rather we say that they have their spins aligned with the field. But it will take only a very feeble energy to turn them around to be pointing in the opposite direction. This we can do by sending radio frequency waves on to them. If we then turn off our transmitter and instead tune in our radio to listen, we will hear (or rather detect by registering the radio signal coming back) the hydrogen atoms spins swing back and align with the magnetic field once again.

For an organic chemist, this is fantastic because hydrogen atoms with different neighbours will "sound" slightly different. Close to an oxygen atoms it might be like a tuba, close to a carbon like a flute, and all together the melody will be a unique manifestation of each molecule. But the pattern can also be analysed like a crossword puzzle to find out the identity of an unknown substance. Few atoms, few words and easy to solve, more complex molecules, more challenging and sometimes real detective work. This we do in small test tubes. These are, however too small to fit humans.

So, from the chemist's nuclear magnetic resonance spectrometer to the physician's magnetic resonance imaging machine, we make two changes. First, we make the magnet big enough so that we can fit an entire person inside, and then we decide to look only at the water molecules. After all, they are the most abundant molecules in our bodies.

Therefore, we will roll you into the tube-formed space inside the MRI magnet and turn on the radio. If the magnetic field has the same strength everywhere it will take exactly the same energy†††† to turn the spins in the opposite direction of the field. When we then listen back, they will all sing us the same tune and we will not be much wiser.

---

††††This is the quantization of energy levels, meaning that only certain energy levels are allowed and, as a consequence, only certain values of energy can be used to push particles or molecules between these levels. This is for example why we have colours, a very pretty and observable effect of quantum chemistry.

There is a trick, however, and that is to make sure that every point in your body will experience a slightly different magnetic field. Now all the water molecules will send us back different notes and if we did not have a computer it would be a real cacophony. Again, it may seem we are not much wiser, many more notes to track, but so what? But the thing is that the water molecules are not equally distributed in your body, in some spots the concentration will be higher, and in some lower. In addition, and even more importantly, depending on the surrounding tissue the rate at which they sing back the radio signal will also vary.

Now in every point in the body we will know which exact note to listen for, and we can then let the computer make us a picture where long-lasting notes are blackish, the longer the blacker, and fast relaxing spins, the term we actually use, are white.

So far so good, and no need for any fancy compounds named after an old Finnish chemist. But the thing is, you would probably like to spend as little time as possible in the MRI machine. Also, many patients need them, and a queue is forming after you as the hospital cannot afford many of these expensive machines. Plus, we need to conserve the helium gas used to cool the superconducting magnetic coils. There is also the issue of keeping still inside the machine, just as with old photographs.

The solution to this is to make the water molecules sing louder in a shorter time, that is, to make them relax back to their original state faster. For this, we need them to interact with something else that behaves like a little magnet, and unpaired electrons are just the ticket for such a contrast agent. The problem is where to find them. Organic molecules normally do not have them, and if so, a so called stable free radical, there will be only one, and more often than not shared over several atoms. To get a good effect, we would like to have several of them concentrated on the same atom.

This is where gadolinium (Gd) comes in, and more specifically the Gd^{3+} ion. This little fellow has seven unpaired electrons in its outer electron shell (and an additional 54 in pairs closer to the nucleus), and it does not get much better than that. Gadolinium is one of the rare-earth elements, and also part of the lanthanoid series, where it sits next to europium (actually found in the Euro notes) and terbium (used in lighting technology) To avoid you

Sr															Y
Ba	La	Ce	Pr	Nd	Pm	Sm	Eu	Gd	Tb	Dy	Ho	Er	Tm	Yb	Lu
Ra	Ac	Th	Pa	U	Np	Pu	Am	Cm	Bk	Cf	Es	Fm	Md	No	Lr

Figure 7.10   Cut-out of the long (32 column) version of the Periodic Table highlighting gadolinium in the centre of the lanthanoid series.

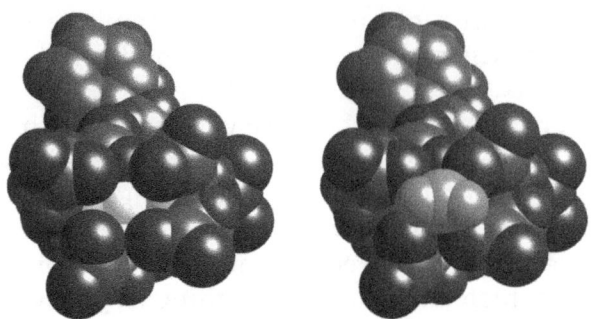

Figure 7.11   In gadolinium contrast agents such as this, the metal ion in the centre is almost completely surrounded by a large organic molecule, leaving a small gap for a water molecule (light grey) from the body fluids to make close contact and 'get relaxed' (*i.e.* drop down from the excited state), as in the right picture. Reproduced with permission from L. Öhrström, *The Last Alchemist in Paris*, Oxford University Press, 2013.

immediately having to look this up in the periodic table, a cut-out is provided in Figure 7.10.

Now we just need the water molecules to get close to the Gd^{3+} ion, then the unpaired electron's magnetic moment will make the hydrogen nucleus relax and emit its radio signal much faster. This we arrange by making the big organic ligand that wraps around the Gd^{3+} ion do this so that there is still space for a water molecule to bind to gadolinium. An example can be seen below in Figure 7.11.

This would not work if the water molecule just glued itself to the Gd^{3+} ion and stayed put, instead it needs to rapidly exchange place with another water so that the maximum number of water molecules are stimulated to relax faster by the gadolinium. It just so happens that the Gd^{3+} ion is also perfect for encouraging this behaviour, the mean lifetime of a water attached to the

Gd^{3+} ion is in the order of a nanosecond. It is not by any means self-evident that it should be this fast. Had it been rhodium 3+ instead the mean lifetime of a water might be up to one and a half years!

Mark Ruffalo got his gadolinium enhanced MRI scan, the tumour could be exactly localised and removed by surgery, and 16 years later he seems to be doing all right, appearing in the 2018 superhero movie *Avengers: Infinity War* as Bruce Banner, the scientist who, when enraged, transforms into the Hulk.

What about the toxicity of the gadolinium ions—should we be concerned about that? Well, here we meet an actor enraged in real life, Chuck Norris. In November 2017 he filed lawsuits against a number of manufactures of gadolinium contrast agents. He claims his wife Gena Norris "was left weak, tired and suffering bouts of pain and burning sensations" after treatment with a gadolinium contrast agent.

Some 300 million doses of various gadolinium complexes have been used since the first one was approved back in the 1980's, and there is certainly no acute toxicity to be blamed on gadolinium. For example, you need almost twice as much gadolinium nitrate as you need potassium nitrate, a food preservative, to kill a rat. We cannot assume that the same applies to all gadolinium compounds though, the big wrap-around organic molecule that engulfs the gadolinium ion may also play a role, and the specific compound it forms with Gd^{3+}. Few problems have come to light so far, although some gadolinium contrast agents are not recommended for people with kidney problems. As with anything "natural" or "synthetic" that we put in our bodies, there is always a case for monitoring short and long-term effects.

Weighing up the pros and cons. Will you be saved from a brain tumour that is threatening your life or that may give you permanent disabilities? Maybe then, most long term effects are acceptable? Having said that, it is good that the scientific community is now looking a bit closer at the long-term effects of gadolinium based contrast agents, as it has recently been found that patients may retain trace amounts of gadolinium in various organs. So far though, nothing alarming has been found, and it is generally believed that the gadolinium compounds are safe, and even safer than other contrast agents used in various other diagnostic procedures.

Seeing and detecting is one thing, doing something about it is another matter. Chromium supplements would not be likely to help celebrated rower Sir Steve Redgrave against his diabetes, as the role of chromium in our glucose management, the starting point of this chapter, seems unclear at best. Indeed, the European Food Safety Authority does not consider "chromium deficiency" as a thing and has proposed no Adequate Intake levels for this metal.‡‡‡‡

There are, however, other metal drugs that may save your life, turn to the next chapter and find out.

REFERENCES

1. T. M. Zydowsky, Of sandwiches and nobel prizes: Robert Burns Woodward, in *Culture of Chemistry: The Best Articles 271 on the Human Side of 20th-Century Chemistry from the Archives of the Chemical Intelligencer*, ed. B. Hargittai and I. Hargittai, Springer, New York, 2015.
2. R. Burks, *Chemistry World*, https://www.chemistryworld.com/opinion/the-strychnine-exhumation/3007131.article.

---

‡‡‡‡"...there is no evidence for the essentiality of Cr(III) in animal nutrition." and "no evidence of beneficial effects associated with chromium intake in healthy subjects" *Scientific Opinion on Dietary Reference Values for chromium*, EFSA Panel on Dietetic Products, Nutrition and Allergies (NDA), *EFSA J.*, 2014, **12**(10), 3845.

Metal Medications and Metals *Versus* Medicines

In the first chapter, we already briefly met the cyclist who cheated both the international sports community and death by using chemistry. Lance Armstrong's recovery from testicular cancer was no miracle, it was to a large extend due to the little molecule cisplatin, more correctly named (short pause for effect) *cis*-diamminodichloroplatinum(II).

But how on earth do you get the idea to use platinum compounds against cancer? It seems a bit medieval to say the least, a remainder from a time when you had better stay away from the doctor, or at least from most of the suggested prescriptions. Mercury against syphilis, for example.

8.1 MERCURY AND ARSENIC

Mercury against syphilis is where it starts. Syphilis was a huge problem, and still is, and the mercury preparations often caused more harm than good. Then, in the years just before the First World War, along came German scientist Paul Ehrlich and the organo-arsenic compounds salvarsan and neo-salvarsan developed in his laboratory. Arsenic is of course another toxic element,

The Rhubarb Connection and Other Revelations: The Everyday World of Metal Ions
By Lars Öhrström and Jacques Covès
© Lars Öhrström and Jacques Covès 2019
Published by the Royal Society of Chemistry, www.rsc.org

but this time the cure worked without doing the patient too much harm, and these molecules remained the standard (and only) treatment until the penicillins came along in the 1940's.[†]

So, strange elements were used in modern medicine even before cisplatin was discovered. Arsenic is sorted into a group of elements known as the metalloids, in which the behaviour might be both that of a metal and a non-metal, the classic example being tin. Tin can change from a metallic form, useful for making buttons and other objects, to a non-metallic form that you can't make much out of if anything at all.[‡]

In contrast to how arsenic behaves in the famous Marsh test, where a metallic silver-like mirror is produced to give evidence of arsenic poisoning,[§] in the salvarsan compounds[¶] arsenic behaves more like a non-metal, but we will not dwell on these compounds further. What we will comment on though, is the way Ehrlich and his team in Frankfurt got to this arsenic compound, the first antibiotic. They did so by synthesising and testing in a systematic way a large number of related arsenic compounds, much in the way that modern pharmaceutical companies work. In stark contrast, this was not what Barnett Rosenberg, Loretta Vancamp, and co-workers at Michigan State University did in the 1960s when they discovered the power of cisplatin.

8.2   THE CISPLATIN DETECTIVE STORY

When Rosenberg and Vancamp chose to pursue an odd-looking phenomenon they had observed in their experiment, they had no idea they were on the track of a block-buster anti-cancer drug. Nor did they know this would change their lives and give a glimpse of hope to thousands of men and women diagnosed with cancer.[1]

[†]Ehrlich got the Nobel Prize in Physiology or Medicine 1908, before salvarsan. Salvarsan is also known as arsphenamine.

[‡]For the myth about "Napoleon's buttons" and the 1812 war against Russia see the chapter on Bonaparte's Bursting Buttons: A Thin Story, in *The Last Alchemist in Paris*, Oxford University Press, 2013.

[§]Agatha Christie learnt the technique as a novice pharmacy student during World War I, and that man of many talents, Lord Peter Wimsey's valet Mervyn Bunter, uses the test in Dorothy Sayers' 1930 novel, *Strong Poison*.

[¶]Neosalvarsan is sodium 3,3'-diamino-4,4'-dihydroxyarsenobenzene-*N*-formaldehyde-sulfoxylate.

What the Michigan State team had originally planned to do was to investigate the effect of electric fields on bacterial cells, but they found themselves instead in pursuit of some unknown molecule formed by the reaction of their electrodes with the water and various salts used in the experiment. These electrodes were supposed to do absolutely nothing other than to be the springboard for the electrons jumping in and out of the water. They were made of the noble and non-reactive metal platinum, periodic table address no. 78, just below nickel and palladium, and just before gold. Common chemical sense suggested that these electrodes should be inert and not react with anything in the water.

To Rosenberg's and Vancamp's surprise, something in the solution stopped the bacterial cells from dividing and producing new cells. This was completely unforeseen, and probably a bit annoying too, but definitely worth looking into, they thought.

The rest of the story reads like a detective novel. Painstaking door-to-door interviews with all neighbours to the "crime": water solution, salt concentration, electric current, electric field, light, temperature and electrode material. All could be eliminated except the combination of platinum electrodes, ammonium ions (NH_4^+, one part of the salt solution), light, and an electric current. By the combination of the salt solution, the electrodes, and the current, a molecule was produced, most likely containing platinum, that was responsible for the biological activity. Now work concentrated on identifying this molecule.

Just as in a good crime story, the team had to dig into the suspect's (that is platinum's) past, spending endless hours in the university library, interrogating long dead scientists through their articles in thick volumes with titles such as *Zeitschrift für anorganische Chemie*. Finally, the team had enough evidence to make an arrest: the suspected compound, diamminetetrachloridoplatinum(IV), $[PtCl_4(NH_3)_2]$, was prepared and tested.

Then, just as in a good suspense story, no effect! They had got it wrong somehow; this was not the guilty compound after all. In a classic twist, it turned out that the suspect was not one person, but two! The compound $[PtCl_4(NH_3)_2]$ was in fact the look-alike siblings *cis*-$[PtCl_4(NH_3)_2]$ and *trans*-$[PtCl_4(NH_3)_2]$. One of them completely innocent with no biological activity, *trans*, and

Figure 8.1   Four molecules with platinum (white) at the centre. From left
to right: the *trans*-[PtCl$_4$(NH$_3$)$_2$] that was inactive, the *cis*-[Pt-
Cl$_4$(NH$_3$)$_2$] that was active in Rosenberg's and Vancamp's tests,
but not developed into a drug, the *trans*-[PtCl$_2$(NH$_3$)$_2$] that was
also inactive and finally *cis*-[PtCl$_2$(NH$_3$)$_2$], better known as cisplatin, the block-buster anticancer drug.

the potent *cis*-brother, *cis* meaning that in this compound the
two ammonia molecules are the closest neighbours (Figure 8.1).
Their first synthesis had only brought the inactive *trans* compound in for questioning.

When the *cis* compound was finally prepared, it tested positive,
and now Rosenberg and his co-workers had a compound that
efficiently prevented cell division, one of the major problems
with cancer cells.

But still much work remained to be done; the painstaking tests
to go from a molecule with biological activity to a useful drug
began, still involving the team from Michigan State University,
but later also the pharmaceutical company Bristol–Myers Squibb.
The first tests on humans were performed in 1971, and in 1978
the drug we now know under the name *cisplatin*, was approved
for worldwide use as an anti-cancer drug.

Before the introduction of *cisplatin*, testicular cancer was more
or less a death sentence, but in 1996 the prospects were better.
Like many other patients, Lance Armstrong was treated with a
mixture of potent drugs, but he took care that the combination
should not give permanent damage to his extraordinary lung
capacity, should he survive.

There are a few additional twists to this story. The drug cisplatin that was finally approved for use is not the *cis*-brother
discussed above, but rather the sister compound *cis*-diam-
minedichloridoplatinum(II), *cis*-[PtCl$_2$(NH$_3$)$_2$]. This compound
has a central platinum atom binding two chlorine ions and two
ammonia molecules in such a way that the assembly resembles

a square, and the NH_3 molecules sit (bind) together on one side of the square, making this compound *cis* in chemists' language.

It is also remarkable that Italian chemist Michele Peyrone from Turin had prepared this compound already in 1844, working in Justus Liebig's laboratory at the University of Gießen,[||] Germany, although at the time the ideas about what this molecule looked like were very sketchy. Following published experimental recipes is not always easy, and Peyrone was not out to find a new platinum compound, he wanted to prepare Magnus' green salt, a combination of the cation $[Pt(NH_3)_4]^{2+}$ and the anion $[PtCl_4]^{2-}$, for further studies of its properties.

Heinrich Gustav Magnus had prepared this compound a few years earlier while working in Jöns Jacob Berzelius' laboratory in Stockholm.[**] Had he conveniently disregarded the yellow crystals formed at the same time as the major green product, or did Peyrone make any changes to the procedure? We do not know, but the funny thing is that the two compounds have exactly the same overall chemical analysis, $PtN_2H_6Cl_2$, (only the proportions could be determined, so $Pt_2N_4H_{12}Cl_4$ is the same thing).

Peyrone scratched his head over this for a long time, until finally admitting the compounds were different, even though they had the same formula. We call such compounds isomers, as did finally Peyrone too, when some years later he was back at the University of Turin. There are many different kinds of isomers and in organic chemistry they are usually formed by arranging the bonds in different ways. But what kind was this? Peyrone had no idea.

It took until 1892, when the French-born Zürich chemist Alfred Werner was able at last to explain what this and the related platinum +IV compounds looked like, and why there were several variations of seemingly the same molecule,[††] and then another 40 years before it was clear why cisplatin had the peculiar flat, square, form and not a tetrahedral structure (*i.e.* tetrapak shaped) as is the case in, for example, many carbon compounds.

[||]Now Justus Liebig University Gießen. And yes, the same guy who was less than civil to the Danish chemist Zeise in a previous chapter.

[**]Located at the Swedish Royal Academy of Sciences, where it had moved from Berzelius' by then rather crowded kitchen in the flat on the corner of Nybrogatan and Riddargatan in Stockholm some years earlier.

[††]Which earned him a Nobel prize in chemistry 1913.

None of these developments were made with the future work of Rosenberg and Vancamp in mind. This was pure curiosity-driven research, but research without which the development of cisplatin would have been impossible.

There is a word for this: serendipity. The Rosenberg team was not just lucky, they discovered something unexpected and acted on that. The French chemist Louis Pasteur expressed it well: "In the fields of observation, chance favours only the mind that is prepared".

8.3   THE IMPORTANCE OF DIGGING DEEP

There is perhaps another thing to learn here as well. With different people and a different scenario, one could envisage the direct promotion of the water solution electrolysed without any additional research. Which is exactly what is going on in the murky waters of colloidal precious metals. These have no documented effects on anything, and in most of the world cannot be sold for human consumption of any kind.

Most notorious is colloidal silver. Once used in desperation to treat infections in early 20th century main stream medicine, as nothing else was available, it was quickly abandoned by physicians and patients alike when penicillin and other modern antibiotics became widely available after World War II. For unclear reasons, and certainly outside the scope of this book, these questionable remedies resurfaced in the 1990's, now rebranded as "alternative" cures for almost anything from cancer to Ebola. To help sell them, the idea of a global conspiracy theory was launched implying that all physicians and all regulatory agencies in Western countries have been bought by "big pharma" ever since the rise of penicillin in the 1940's.

Let's be clear on a few things; silver metal is mostly harmless and is an approved food colouring additive in the European Union. Silver 1+ ions, Ag^+, are good or moderate antimicrobial agents when used in test-tube assays, and for disinfecting drinking water there may be some sensible applications. However, the innocence of Ag^+ ions may be overstated. NASA has been using silver in the potable water system aboard the Space Station ISS, but there have been some recent concerns: "...studies have shown the possible toxicity of colloidal silver to humans, including crew members aboard the ISS".[2]

Inside the body though, it's a different story, and so far, silver preparations have proved mostly useless against internal infections and other illnesses. The difference between killing a few germs in otherwise mostly clean water and the vicious chemical soup contained in the stomach with its hydrochloric acid is like the difference between the rain forests of the Amazonas and the plains of Antarctica.

On the contrary, in wound healing applications there are some, but not very strong, indications that bandages with different silver contents (including "nano" preparations) may somewhat speed up the healing process and help prevent infections. This might seem like a small issue, but wound healing and its complications consume significant resources in our hospitals, and cause considerable suffering, especially for large diabetic-related difficult-to-heal wounds, and large burns. As the latter might be life-threatening should an infection occur, even weak indications of a positive effect may be reason enough to use such medical devices.

Even though silver ions have a low toxicity for humans and mammals in general, they might nevertheless be harmful to the environment. Therefore, even if colloidal silver might, and we stress "might", be mostly harmless in low quantities, it is completely unnecessary to pollute our waters and bodies with this old fashioned and, as far as we know, useless remedy. As for other ways of using the antibacterial properties of silver, ongoing research needs to provide the answer.

To summarise, had the researchers working on the platinum electrolysis taken the easy way out and just marketed some kind of platinum solution, thousands of lives might have been lost every year because cisplatin and its cousins may never have been developed.

8.4   THE TIRESOME MINGLING PARTY

We are now fairly sure that cisplatin, and the three or four analogous drugs, work by first passing through the cancer cell's membrane as neutral molecules. This is important, as ions can only pass into the cells through special selective tunnels or by being picked up by the taxi drivers of the body, the transport proteins. These are, however, extremely picky when choosing customers so not just any ions can hitch a ride with them.

Then, to do its job, cisplatin needs to get to the nucleus and the DNA of the cancer cells to stop them splitting into new cells, growing the tumour or invading new parts of the body. This requires a slight change of costume. DNA is a long polymer in which the units are held together by phosphate ions, PO_4^{3-}. Up and down the helix they connect sugar units, each having a charge of plus one, and each sugar unit binding one DNA-base, the information units. This means there is an overall negative charge on the whole DNA polymer because of the phosphate groups.

This will attract positive ions, but cisplatin is neutral. Luckily, however, the picture we have painted of a rigid molecule with strong bonds is slightly misleading. The bonds are fairly strong, true, but this is not the whole story. We come much closer to the truth if we think about a tiresome mingling party.

Many guests have been invited but there are only a few square tables with four seats at each, and with about as much space to occasionally make a fifth person fit, although then everybody becomes rather uncomfortable. Two people are permanently seated next to each other at each table, the privilege of old age perhaps, whereas the other two seats may be occupied by anyone in the party. When a fifth guest arrives and temporarily becomes part of the conversation, it quickly gets too crowed and uncomfortable, so that one of the three exchangeable guests will quickly leave, restoring the foursome, but perhaps with one of the guests swapped for the newcomer.

At first the tables are set outside, and most guests have jackets (we can imagine this is in Sweden), and after a while of swapping around, this means that at most tables the exchangeable guests will have jackets. When the table is moved inside, the scene changes and most people head for the cloakroom to dispose of their jackets. Now we will soon see jacket-less people seated at most tables, and this is more or less what happens when the cisplatin moves from outside the cell, with a high concentration of chloride ions (jackets on) to inside the cell where there is a low chloride concentration and many more people without jackets (water molecules). The seating arrangements also happen in the same way; a fifth ligand temporarily attaches

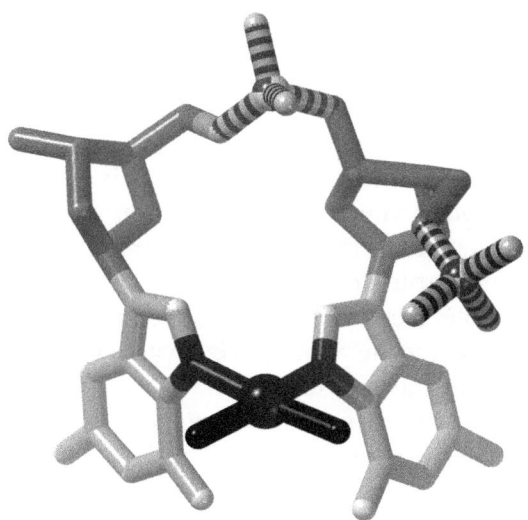

Figure 8.2   Cisplatin, having lost two negatively charged chloride ions after entering the cancer cell, can now be attracted to the negatively charged DNA helix and bind to two neighbouring bases causing a serious hitch when the DNA unzips in the cell division process. With a bit of luck, it will stop completely and the cancer stop growing and eventually disappear. In this picture we have omitted all hydrogen atoms and the square-planar platinum unit is shown in black, the DNA bases in white, the sugar units attached to the DNA bases are shown in grey, and finally the phosphate ions that hold the DNA helix together are shown with striped bonds.

to the platinum giving a high-energy molecule that quickly goes back to a lower energy state by detaching one of the exchangeable ligands.[‡‡]

Now, suddenly we have a cation, $[Pt(NH_3)_2Cl(H_2O)]^+$, that will be attracted by the negative DNA helix. Once close, the many nitrogen atoms in the DNA bases become easy targets for the platinum cations. The water molecule or chloride ion will be replaced in much the same way as before, first with one, and then with a second base, a neighbour to the first and conveniently close. And now the platinum ion is firmly anchored to the DNA helix (Figure 8.2).

[‡‡]The main difference is that in the case of the molecule, the new guest comes in from the "top" or the "bottom" of the square, not from the side. Although we might occasionally have guests dancing on the tables, we hope we can be excused for this over-simplification. Also, we should add that for metal ions with six-coordination, the exchange might take place both in the "associate" way just described when sitting around the tables, but also by one of the guests just moving on and leaving an empty seat behind, a "dissociative" mechanism.

8.5   UNZIPPING THE DNA

The DNA double-helix is an iconic picture of modern science, both because of its striking appearance and the easy-to-draw images that can be created, but also for how its discovery was stolen from Rosalind Franklin by Francis Crick and James Watson with Franklin's boss Maurice Wilkins as a, perhaps, unwitting co-conspirator. A gamble that sort of misfired for Wilkins, as had he and Franklin published the results together, Wilkins could have been the sole receiver of the Nobel Prize in Physiology or Medicine, because by the time a prize was awarded for the DNA structure, Franklin was already dead.

Instead, Wilkins first became the one-step-behind Nobel laureate of 1962 that nobody remembers, and later the willing accomplice handing the unscrupulous Crick and Watson fame and fortune on an X-ray photographic plate. Crisply described in the play *Photograph 51* by Anna Ziegler, this 2015 West End production was noted for a critically acclaimed performance by Nicole Kidman as Rosalind Franklin.[§§]

But we are digressing. What we mean to say is that, iconic as it may be, the double helix doesn't tell the whole story, we also need the zipper, because it tells us more about the function when the cell divides. This of course is the big problem with cancer, the rapid cell division, and for it to proceed, a copy needs to be made of the cancer cells' DNA.

This happens by gradually unzipping the DNA double helix to two separate helices and, subsequently, with the help of enzymes, building two new copies of each strand by adding, piece by piece, the complimentary DNA bases until there are two identical zippers, each with one half from the original double helix. When this process is completed, the new cell can have its own identical copy of the DNA. However, when there is a platinum ion hanging on the side, it will create a kink that will disrupt this fine-tuned molecular machinery. We are all familiar with uncooperative zippers so we know how critical this can be, and the DNA-zipper-kink efficiently stops the cancer from growing.

Cisplatin also has an effect on other cells in the body, but the rapidly multiplying cancer cells will be hit much worse than our

---

[§§]Incidentally, Kidman's father was a biochemist.

normal cells. Therefore, Lance Armstrong was cured, and the sales, in the order of many hundred million dollars a year, have given a good income to Michigan State University *via* the license agreement with pharmaceutical company Bristol–Myers Squibb. Armstrong subsequently appeared in many commercials for Bristol–Myers Squibb, but some think this company has taken a bit too much credit for a drug that had a very long academic history.

8.6 AROUND THE NEXT BEND

The development of cisplatin was carried out in the 1970's, and if we are still waiting for the next breakthrough in metal-ion-based anticancer drugs, it is not for the want of trying. Early attempts included making cisplatin-lookalikes by simply substituting platinum for palladium. This compound is also square planar but failed because the accidentally well-timed process in which cisplatin changes its costume twice on its way to the DNA is much faster with palladium. The $[PdCl(NH_3)_2(H_2O)]^+$ ion is simply formed much too early to have a chance of ever reaching the nucleus before being attacked and destroyed by other chemicals in the complicated cell-soup.

We are not going to give you an extensive review of what is going on in this field, as it is a lot, mainly in academia, where cisplatin had its origin, and not so much in pharmaceutical companies. To the annoyance of some that feel the industry is a bit too conservative.

So here is just a glimpse of a type of compound that just might one day make it to the shelves of the pharmacy. In Texas chemistry professor Kim Dunbar's lab, dirhodium complexes are being investigated as photochemotherapy agents against cancer (Figure 8.3).

8.7 I DON'T REALLY BELIEVE IN GOD, BUT I BELIEVE IN LITHIUM

A battle scene in the 2017 *Star Wars: The Last Jedi* movie takes place on the planet Crait. This fictional planet is too far away even for a movie crew backed by Lucasfilm and Walt Disney, so the very secretive filming instead took place in a, for chemists, even more

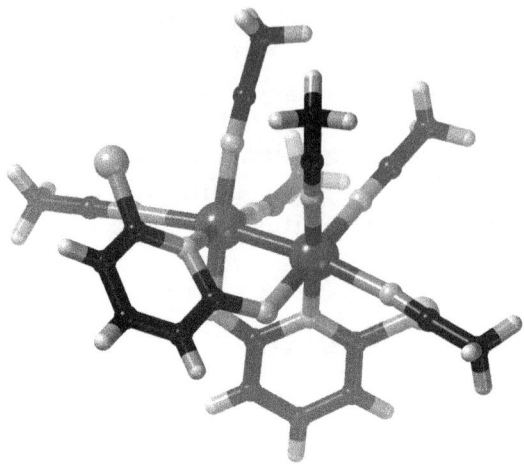

Figure 8.3   Professor Kim Dunbar has a long track-record in working on anti-cancer compounds. This di-rhodium compound is just one example of the compounds coming out of her laboratory at Texas A&M University.

interesting place, the World's largest salt pan, Salar de Uyuni in Bolivia. Covering an area of 10 582 square kilometres, the size of Lebanon, it is remarkable also for hosting one of the world's largest reserves of lithium in the form of lithium chloride, LiCl.¶¶ This salt closely resembles sodium chloride (NaCl), potassium chloride (KCl) and magnesium chloride (MgCl₂), the other major components of the meter thick brilliantly white crust.

Does the atmosphere in the Salar de Uyuni have a calming effect on the tourists that come from all over the world? Perhaps it is difficult to distinguish from the sense of wonder at seeing this unique place. But lithium ions, Li⁺, are indeed a common treatment for bipolar disorder. Lithium carbonate, lithium citrate and other lithium salts cut off the manic phase of the condition, not so long ago known as manic depression, and have helped many people to live normal lives (Figure 8.4).

¶¶The climax of the James Bond movie *Quantum of Solace* (2008) was filmed in the neighbouring Atacama Desert in Chile, another place with remarkable lithium reserves. The evil plan that Bond needs to thwart this time has to do with water, but given the increasingly strategic importance of lithium, most of it used for our ubiquitous lithium ion batteries, element three might have been a more logical choice for a plot device as we get most of our lithium from South America.

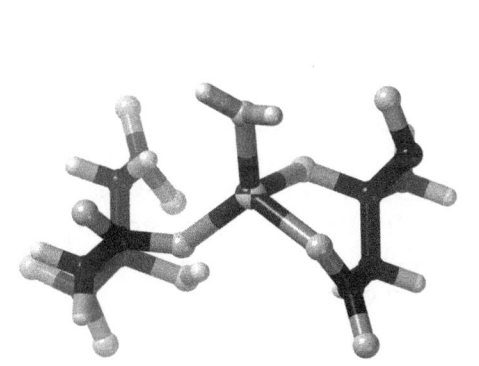 

Figure 8.4 Left: Part of the solid-state structure of lithium (beach ball) citrate.[IIII] In the body, the lithium ions will mostly separate from the citrate anions (from citric acid, the common sour ingredient in all sorts of citrus fruit) and form hydrated lithium ions, that is $[Li(H_2O)_4]^+$. Right: In a time when the difference between soft drinks such as Coca-Cola, and patented medicines with dubious effects, was fuzzy, or perhaps fizzy, there was something in the US called "Bib-Label Lithiated Lemon-Lime Soda" formulated in 1929 and containing lithium citrate. Supposedly good for hangovers, the drink later became known as 7Up and lithium citrate was removed in 1948. Incidentally, the most common isotope of lithium, making up 92% of all naturally occurring lithium atoms, is ^7Li with the 7 in the "up" (superscript) position, and the atomic weight is therefore close to 7. Reproduced from https://commons.wikimedia.org/wiki/File:Old_7UP_bottles.jpg under the terms of the CC BY 2.0 license, https://creativecommons.org/licenses/by/2.0/deed.en.

For some, like British actor Stephen Fry, who was diagnosed with bipolar disorder in early middle age, the choice was not clear. He considered the possible side effects, but was also wondering if the manic phases had perhaps helped him in his career, and if so, would he lose something valuable in taking the drug?

For Jaime Lowe, there was never such a choice. The manic phases she tells us about in her 2017 book *Mental: Lithium, Love and Losing My Mind* are truly fearful. At 16, she believed

---

[IIII]This compound is a polymer in the solid state, every oxygen atom connects to a lithium ion, the water oxygen atoms to two, and the formula is $[Li_3(citrate)(H_2O)]$. We will see some similar compounds in a coming chapter.

her parents were evil secret agents and that war was imminent. She was admitted to the emergency room at a Los Angeles hospital where she was eventually diagnosed with bipolar disorder. Lithium then helped her live a normal life for the next 20 years, to the extent that she declared: "I don't really believe in God, but I believe in lithium".

Side effect are real though, and constant lithium exposure has taken its toll on her kidneys, so after 20 years she needs to get off it and on to something else. Which turns out to be something of a struggle and forms the final part of her book.

8.8   COSTUMES, SIZES AND A CLOSE ATTRACTION

How exactly lithium works on a biochemical level is still actively researched. It has no known "natural" role in the body, so we are not talking about some kind of lithium deficiency. Much remains to be done, and the ultimate goal is of course to see if we can make the present drugs better, avoiding the kidney damage experienced by Jaime Lowe for example.

One line of inquiry is that Li^+ may work on the ion channels in our neural signalling system, replacing sodium ions and calming it down. This may be down to its reluctance to change costume. The lithium ion is very small, the smallest metal ion we know, with a +1 charge,*** and is substantially smaller than the sodium ion, Na^+. Bizarrely, the way our body sees things, this makes the Li^+ ion larger because its smallness and "hardness" makes it hold the surrounding water molecules in a tight grip, much tighter than sodium. This is because the negatively charged oxygen end of a water molecule comes much closer to the lithium nucleus than the sodium nucleus. This means the positive charge comes closer to the negative and the bond therefore becomes stronger.††† So while Na^+ ions may easily strip off their waters of hydration and escape, appearing almost naked,

---

***It has atomic number three, and consists of only three protons, four neutrons (for the most common isotope lithium-7) and three electrons. Only the elements hydrogen and helium have lower atomic weights.

†††Of course, there is a formula showing this; the energy of interaction between a dipole (a molecule having positive and negative ends, such as water) and a charge is proportional to (charge) × (dipole 'strength')/(square of the distance), or as we prefer to write: $E = k \cdot q \cdot \alpha / r^2$.

the lithium ions will wear a heavy overcoat of water molecules no matter what the weather and will have more of a struggle to get through the ion channels.

8.9 THE SARCOPHAGUS MOLECULE

Chemist sometimes struggle with systematic names and their uses but are sometimes quite inventive. What to call a molecule that will completely encircle a metal ion, giving it no way to escape? Well, the molecule formally known as 3,6,10,13,16,19-hexazabicyclo[6.6.6]icosane is in common chemist's language simply called sarcophagine, because once you get a metal ion into that cage, it will have a hard time getting out again (Figure 8.5).

Sarcophagines loaded with ^{64}Cu or gallium-68 (^{68}Ga),[‡‡‡] are a type of molecule that finds use in one of the most cross-disciplinary fields there is, nuclear medicine. Nuclear medicine is not

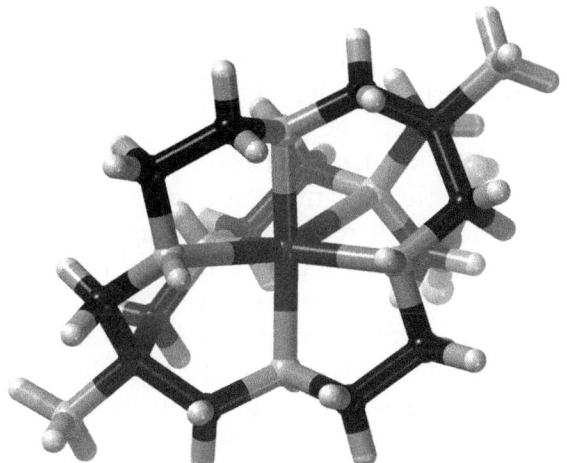

Figure 8.5 Some ligands form a pre-shaped cage with no need to fold, as is required of the EDTA molecule. They are commonly known as sarcophagines because what's inside the coffin finds it very difficult to get out, just like someone in an Edgar Allan Poe story, or like Cu^{2+} in this case. Formally these ligands are known as being based on 3,6,10,13,16,19-hexazabicyclo[6.6.6]icosane.

[‡‡‡]Gallium-68 and ^{68}Ga are synonyms, but the former is easier to say and therefor better suited to a text that is to be read out.

about assessing and curing damages from ionizing radiation, it is about using radioactive isotopes of a large variety of chemical elements in medical diagnosis and occasionally treatment (sometimes called theranostics). Radiation safety is of course one important aspect (in Chapter 5 we briefly mentioned the Goiânia accident, in which a medical device containing radioactive caesium went missing), but the field also combines medicine, biochemistry, organic and inorganic chemistry, nuclear chemistry and physics.

Earlier in this book we talked about seeing inside the body and have already encountered technetium in the previous chapter. Other radioactive isotopes routinely used in hospitals are ^{11}C, ^{13}N, ^{15}O, ^{18}F, ^{89}Zr (zirconium-89), and ^{82}Rb (rubidium-82) and they are tracked in the body by positron emission tomography, the most commonly device and better known as just PET, or as doing a PET-scan.

The beauty of the method is that you can use extremely small quantities of the radioactive materials, roughly you need concentrations such as if you dissolve a couple of grains of sugar in one of the Olympic pools from London 2012. This works because we can in principle detect single atoms by the radiation they send out.

In the case of a PET scan, these isotopes break down by making the unstable atomic nucleus more stable by converting a proton to a neutron, which both reduces the proton–proton repulsion and adds more padding to screen the other protons from each other. The charge must still balance though, so a positron (an electron but with a positive charge) is simultaneously sent out from the nucleus. The positron is an anti-particle and it will not travel far before it hits a normal electron. When this happens, both particles disappear (as this is an annihilation reaction between matter and antimatter) and the energy is converted to two gamma rays that are sent out with an angle of $180 \pm 0.5°$ from each other. Having a detector all around the body means that we can, with the help of a computer, calculate exactly where in the body this "event" has taken place (well, within some millimetres anyway).

To make the whole procedure useful, these isotopes need to go where the tumours are, or where we need to see other things.

For this, "labelled" organic compounds are also very common, and the PET scans became really popular only after Johanna Fowler and her colleagues at Brookhaven National Laboratory in the US had synthesized ^{18}F-fluorodeoxyglucose (FDG) in 1976.[§§§] This molecule resembles normal glucose very much, only one hydroxyl group (–OH) has been exchanged for one fluorine atom, and it goes wherever glucose goes in the body. By doing that it has been most helpful in imaging the brain, and helping physicians fight cancer.

8.10 WE MIGHT SEE MORE OF PARIS

These compounds do not merely act as scouts though, there are also ways to fight cancer with them. If we choose an isotope that sends out radiation than can kill cancer cells instead of just registering where they are, we have a potential drug. We just need to anchor a molecule that will recognise a cancer cell to a molecule that will make sure the radioactive isotope doesn't go wandering about and do damage in other places, and then, finally add the right isotope.

It turns out that lutetium-177 or ^{177}Lu might be just the ticket. Never heard of this element? No wonder, it is rarely used and not very abundant either. To add to its misery, it is usually placed in a remote corner of the periodic table, at the very end of the lanthanoid series.[¶¶¶] However, appropriately for what we are going to tell you now, element 71 was, after some international commotion, named after Lutetia, the Roman name for what was later to become the City of Light: Paris.

It is not unlikely that we might hear more of this element in the future, as it is right now being successfully used in experimental and clinical treatments against some severe cancers.

[§§§]Meaning that one of the atoms in a molecule has been replaced by a specific, often radioactive, isotope of the same element. A tricky procedure often requiring very clever chemistry. Fowler received the National Medal of Science from President Barack Obama in 2009.

[¶¶¶]With its valence electron configuration [Xe]4f^{14}6s^{2}5d^1, element 71 seems to belong to group 3, and its downstairs neighbour lawrencium, Lr, for which experimental data are much harder to obtain, is in the same ambiguous situation. Chemists have debated for some time now which elements that should be placed below scandium (Sc) and yttrium (Y) — lutetium and lawrencium, or lanthanum (La) and actinium (Ac)?

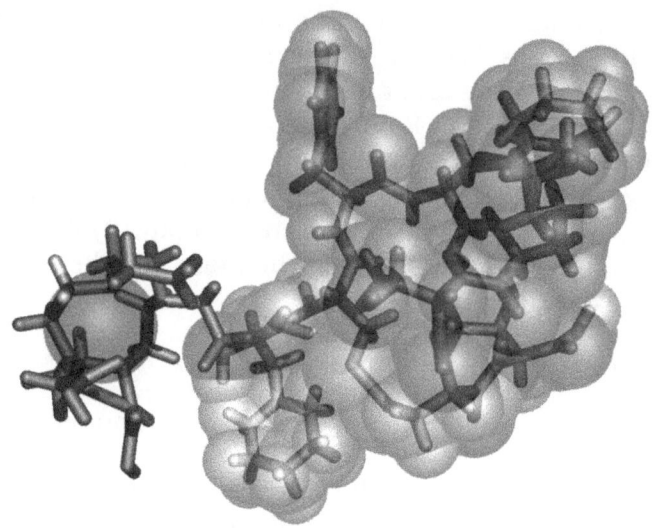

Figure 8.6 A radiolabelled somatostatin analogue with a DOTA ligand derivative enclosing lutetium (left part); the part of octreotate that binds to the receptor is to the right. Image credit: Dr Britta Langen, University of Gothenburg, Sweden.****

By hooking Lu^{3+} up to a tetraaza-cyclododecane-tetraacetate ligand (DOTA) grafted to octreotate, a small peptide, we get a tiny guided missile.‖‖‖ DOTA is a seven or eight-coordinating chelator, which is needed for this large cation, while the octreotate peptide binds to receptors on the surfaces of a number of different types of tumour cells (not all cancers are the same), thus directing the ionizing radiation of the lutetium isotope into the tumour and killing it (Figure 8.6).

‖‖‖The method is also known as targeted radionuclide therapy, TRT. ^{177}Lu can be produced by neutron radiation of the natural isotope ^{176}Lu by just adding a neutron, 176 + 1 = 177. Alternatively, by neutron radiation of the likewise natural ^{176}Yb, giving ^{177}Yb in the same way. ^{177}Yb will after a few hours have changed a neutron to a proton, the opposite of what happens in the positron camera reaction, thus going from element 70 to 71, and in the process sending out an electron (β^- decay) to preserve the charge neutrality, giving ^{177}Lu.

****More formally: ^{177}Lu-DOTA0-Tyr3-octreotate showing a potential conformation built using Protein Data Bank structures 1YL8.pdb (Tyr3-octreotate) and 1NC2.pdb (adapted DOTA) using PyMOL (The PyMOL Molecular Graphics System, Schrödinger, LLC).

Just like a sarcophagine, this ligand holds the metal ion in a tight grip, just like in an Edgar Allan Poe novel there is no escape out of the coffin.

8.11 DRUG TRAFFICKING ACROSS THE PACIFIC AND ACROSS MEMBRANES

In 1949, a young woman in Australia had her life saved in an unusual way. She is critically ill with pneumonia at a Melbourne hospital but her father has been told perhaps she could be saved by a new wonder drug, aureomycin. The problem is, you cannot get aureomycin in Australia yet. Her father, however, is a keen amateur radio operator and gets talking to a fellow amateur in the US, who happens also to be a professor, and mentions the problem. Suddenly, they get interrupted by a third operator that has been listening in, a pharmacist in Kalamazoo, Michigan, who offers to help, and the next day a sufficient sample of the drug is on its way to Australia on a Pan Am flight (Figure 8.7).

This was just two years after substantial amounts aureomycin, the first of the tetracycline class of antibiotics, had been produced. This drug was developed under the direction of Yellapragada Subbarow, a very talented chemist from Andhra Pradesh in India. After completing a PhD thesis at Harvard, where amongst other things he described the importance of ATP, the universal chemical energy currency of all living things, he went on to a position in the Lederle Laboratories in Pearl

NEW DRUG SAVED GIRL'S LIFE AFTER AIR TALK TO US

*S*YDNEY, Tues: A North Sydney amateur radio operator tonight wirelessed amateur radio operators in the United States that a drug they had sent to Australia had saved his daughter's life.

Figure 8.7   Headline in "The Argus", Melbourne, Australia, 7 September 1949. The Australian drama TV series Flying Doctors, 1986–1993, was hugely popular even overseas, but this news is about a new wonder drug being flown into Melbourne from the USA.

River, New York, where even more remarkable things were to happen.[††††]

Followed by the determination of the chemical structure of folic acid, a story that began with Lucy Wills in the UK, a lady we will hear more about in the next chapter, Subbarow's team came up with methotrexate, an anti-cancer drug we still use, and *via* war work on penicillin, they discovered aureomycin.

The activity of aureomycin was noticed in September 1945 when samples from the growth of a bacteria living in soil were tested against some nasty bugs. But making large amounts of the new potential drug turned out to be difficult. It was not until February 1947 that calcium ions finally came to the rescue.

The bacteria tended to die before producing much aureomycin, and as these molecules were potent antibiotics (bacteria killers) this was perhaps not too surprising, but there was also the matter of too much acid being produced. The solution turned out to be adding the common constituents in limestone and marble: calcium carbonate, $CaCO_3$, thereby killing two birds with one stone. The carbonate ions took care of the excess acid by making bicarbonate ions (HCO_3^- as in baking powder) and then the calcium ions formed an insoluble compound with aureomycin that crashed out onto the bottom of the reactor, leaving the bacteria in peace to produce more.

But what rescued the project at this stage is also a problem if you are on tetracycline therapy. Avoid milk when taking the drug, the calcium ions will make the drug insoluble in the same way and it will just pass right through your stomach and out of the other end (Figure 8.8). Once digested, however, it seems metal ions such as magnesium 2+ take on a more beneficial role and might be essential for both the transportation across membranes, and the work that needs to be done: to block the bacteria's synthesis of new proteins.

8.12   WRAP AROUND AND DELIVER

Wrapping this chapter up, we will revisit cisplatin very quickly. One of the first patients to be cured by cisplatin was newly-wed 23 year-old John Cleland in 1974, some years before the

----

[††††]ATP is adenosine triphosphate, and this and other achievements at Harvard has made Siddhartha Mukherjee suggest Subbarow was denied a permanent position (tenure) because he was "a foreigner, reclusive, nocturnal, heavily accented vegetarian", *The Emperor of All Maladies: A Biography of Cancer*, Scribner, 2010. Lederle laboratories was at the time part of the American Cyanamid which, at the time of writing, belongs to Pfizer.

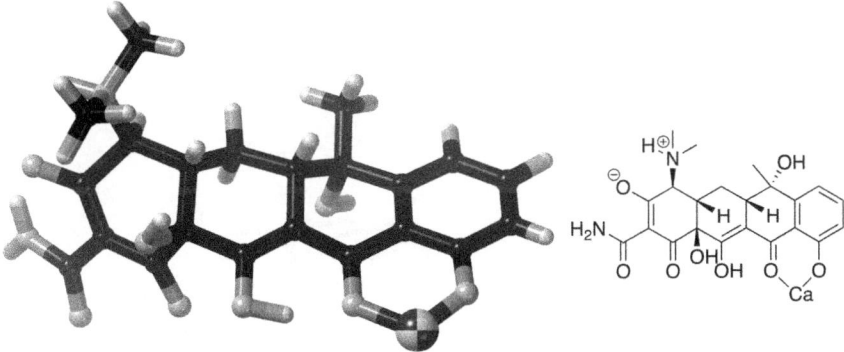

Figure 8.8 A tetracycline molecule (aureomycin in this case) having trapped a calcium 2+ ion (beach ball). This was the key to making enough of this new antibiotic in 1947 and is also the reason you should avoid metal ions, such as calcium in milk, when you are using tetracycline treatments. These compounds are insoluble!

drug was officially available in the US. Cleland had testicular cancer that had spread to the lungs, and a bleak, and very brief, future lay ahead of him. Luckily, his physician Lawrence Einhorn also did research at the Indiana University Hospital where Cleland was treated, and decided to offer him a last resort gamble, after the failure of three earlier chemotherapy regimens. He suggested a combination of cisplatin, still four years from approval by the Federal drug Agency, FDA, in the US, and the known anticancer drug rose periwinkle.[‡‡‡‡] To this he also added another new agent approved just the year before, bleomycin.

Bleomycin is natural product discovered in 1966 when the Japanese scientist Hamao Umezawa got lucky when he fished for biological activity in various fermentation broths. No metal ions in this molecule, but to do its job, it seems that bleomycin needs to pick up a metal ion on its way to chop up the DNA of the cancer cells. The thing with bleomycin is that it wraps around Fe^{3+} but only occupies five of the six positions with the nitrogen atoms it has available (Figure 8.9). This leaves one position open, often a requisite for catalysis, and oxygen, O_2, can if attached here, generate highly reactive radicals causing oxidative degradation of the DNA, which is what kills the cancer cell.

[‡‡‡‡]*Catharanthus roseus,* also known as the Madagascar periwinkle, is traditionally used in Indian and Chinese medicine.

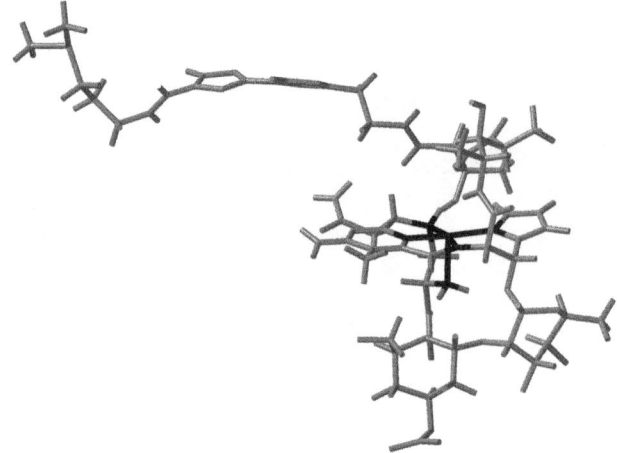

Figure 8.9 The bleomycin molecule (grey) wrapping around an iron ion, grabbing it with five nitrogen atoms (black). In the empty slot above iron, oxygen in the form of O_2 can come and activate the molecule for DNA destruction.

Cleland survived, and as he recollects 40 years later; three weeks after the start of the combination treatment Dr Einhorn came back with some newly developed X-ray plates and declared, "John, the chest X-ray is clear. I think you're gonna make it!"[§§§§]

However, platinum is not an essential metal to have in your body, so you should stay away from it unless ordered by your doctor. In the next chapter though, we will learn of some metals that you'd better not miss out on if you want to stay healthy.

REFERENCES

1. J. Wennerberg, *Läkemedel som förändrat världen: historier om vetenskap, slump och envishet*, Apotekarsocieteten, Stockholm, 2012.
2. International Space Station Internal Environments (ISS Internal Environments) - 08.16.17, NASA. https://www.nasa.gov/mission_pages/station/research/experiments/1060.html.

[§§§§]Patient and physician were reunited and interviewed 40 years later for The Cancer Letter. Matthew Bin Han Ong, "40 Years Later: Doctor and Patient Reflect on the Cure", Oct. 24, 2014, https://cancerletter.com/articles/20141024_1/, read 17 Jan. 2018. Combination chemotherapies like these are standard treatments nowadays.

CHAPTER 9

Blood, Iron and Body Metals

On Gray's Inn Road, on the way to King's Cross railway station from the Charles Dicken's museum in central London, lies the buildings of the former Royal Free Hospital. Here, in 1944, Lucy Wills in the pathology department was working on the problems of anaemia, while Gladys Hill was running the maternity unit as best as she could under wartime conditions.

The war, among other things, meant strict rationing of food. Which was hard, but the efforts to establish fair rations had, through the provision of special allowances for pregnant women, unexpectedly "improved the diet of expectant mothers". The problem of anaemia, a low count of red blood cells, low levels of the oxygen catching protein haemoglobin and low ability of the blood to carry oxygen, still persisted during pregnancy though. Wills and Hill, with an all-female team, therefore designed an experiment to see if supplementary iron in the form of iron(II) carbonate would help (Figure 9.1).[†]

[†]They used approximately 580 mg $FeCO_3$ per day in the form of so called "Blaud's capsules", equivalent to 280 mg elemental iron. Today WHO recommends supplementation with 30 mg to 60 mg of elemental iron and 400 µg (0.4 mg) folic acid for pregnant women to prevent maternal anaemia, puerperal sepsis, low birth weight, and preterm birth. See http://www.who.int/elena/titles/guidance_summaries/daily_iron_pregnancy/en/. Before the molecular identity of folic acid was discovered, briefly mentioned in the previous chapter, it was known as the "Wills' factor".

The Rhubarb Connection and Other Revelations: The Everyday World of Metal Ions
By Lars Öhrström and Jacques Covès
© Lars Öhrström and Jacques Covès 2019
Published by the Royal Society of Chemistry, www.rsc.org

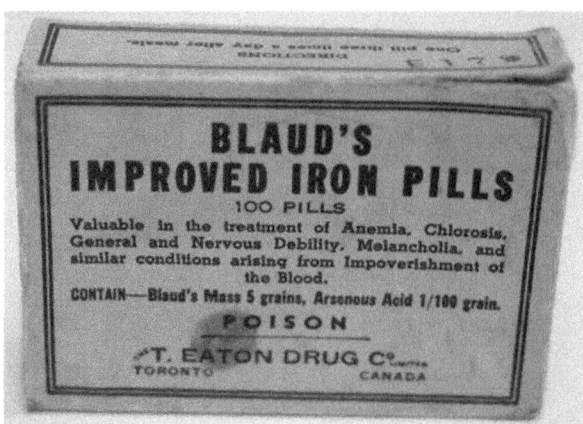

Figure 9.1   In Victorian times, women were seen as the weaker sex, needing all
sorts of strange supplements and tonics. Iron supplements were
a good idea though, introduced by the French physician Jean-
Pierre Blaud (1773–1859) in the small town of Beaucaire on the
Rhône in the south of France. But iron in combination with arsenic, as seen in this preparation, cannot be recommended. This
picture [1015-T17] is in the Medical Instrument and Pharmacy
Collection #HC0013, Health Sciences Library, University of North
Carolina at Chapel Hill.

9.1   IT DID NOT REQUIRE A SHERLOCK HOLMES

The set-up was the classical double-blind method, identical capsules with or without iron were prepared and given to the moth-
ers-to-be and only the investigators could look up the records
and see who was getting what. Thorough analysis of blood samples was then performed, but the double-blind method soon lost
its significance as the all-women team writes in their 1947 report
in *British Journal of Nutrition*: "The laboratory workers were not
told which patients were receiving iron, but after a few visits it
did not require a Sherlock Holmes to ascertain the nature of the
capsules being taken."

That the study was a success, despite wartime hardships
including a direct hit by a German V1 rocket, laconically reported
in the scientific paper as "Unfortunately, a flying bomb incident
interrupted the work", was probably much down to the second

author Gladys Hill,[‡] who managed clinical care at the London Free throughout the war.

Iron deficiency, together with iodine deficiency, continue to be huge public health problems around the world, and this study was an important stepping stone in understanding how it could be addressed. We will use it as a way into the human body to investigate the role of metals in our biochemistry. It is not a comprehensive presentation, or our editor would have gone mad. One third of our proteins use a metal ion in some important way, and there are about 20 000 different proteins that we use at one time or the other in the different cells of our body, so you would have been holding a much thicker book in your hands in that case.

We will instead pick some examples that we find important, and we start with taking your first breath as a new-born baby, getting oxygen from the air to stick to the heme molecules in the haemoglobin protein present in our red blood cells (there are four hemes in each haemoglobin). The blood then flows out to your muscles and there the oxygen molecule will be transferred to another protein, myoglobin,[§] with just one heme group.

As we are all born equal, our haemoglobin proteins are the same, colouring our blood red, not blue, with a slightly different shade depending on if there is a dioxygen molecule sitting on the iron (bright red) or not (dark red). In other species, the red colour looks much the same, but the haemoglobins will differ in small molecular details such as having specific sugar molecules attached. These differences make direct use of pig's blood, or indeed blood of a different human blood group, impossible to use in blood transfers, but knowing the chemistry of haemoglobins, we can possibly put this to use in other ways. And perhaps a can of worms could help?

[‡]Gladys Hill (1894–1998) worked for MI5 during World War I with "clerical duties and passport control" which might mean just that, or something completely different... She later studied medicine in London and the medical use of radium under Marie Curie in Paris. She got her MD in 1925 and started work at the "Royal Free" in 1940. See https://livesonline.rcseng.ac.uk/biogs/E008666b.htm, The Royal College of Surgeons of England.

[§]In general, "globin" in a protein name means that it has a spherical shape.

9.2 MODERN COLOUR-BLIND VAMPIRES

On the sandy beaches of the Atlantic Ocean at low tide you can see the coiled castings of the polychaete annelid *Arenicola marina*, better known as the lugworm (or sandworm). This large worm, fully grown species are around 20 centimetres long, has a peculiar type of haemoglobin with an ability to transport oxygen that is 40 times higher than that of human haemoglobin. Moreover, it does not need any red blood cells to function, so it works as a stand-alone and very big molecule in solution. It is also anonymous, lacking the soldier's uniform made up of a surface decorated with specific sugar molecules used for identification. This is important, as otherwise our immune defence would go berserk when a foreign protein was detected. Instead, this incognito oxygen carrier can slip in undetected and do some good work, for example as a blood substitute in the case of cut blood supplies, as an alternative to a blood transfusion in case of emergency (battlefield for instance), and for organ preservation prior to transplantation.

In these modern and complicated times, in which we are not far off being recommended to get our protein load by eating crunchy insects, you can even get humanized pig kidney transplants if a compatible donor is not immediately available. Graft survival is a crucial problem though, and apart from the immunologic compatibility, the time window of transfer between the donor and the recipient of a transplant is very tight as the organ to be grafted must be constantly oxygenated. A method to increase the preservation time is to immerse the organ in a blood substitute based on some non-human kind of haemoglobin, such as that of the sand worm. The fact that it is a natural molecule allows its production in large amounts by genetic engineering of bacteria involving recombinant DNA techniques, avoiding possible shortages of sea worms in the future.

The fact that this may be a serious issue is shown by the case of the horseshoe crab, *Limulus polyphemus* (Figure 9.2). In contrast to the unfortunate king of France Louis XVI, who was guillotined on Monday 21 January 1793 in Paris, the blood that flows in this arthropod is blue, as expected for an aristocratic member of the nobility, but severely lacking for King Louis.

Figure 9.2   The blue-blooded horseshoe crab, *Limulus polyphemus*, painting by Heinrich Harder, *c.* 1916, public domain.

The horseshoe crab is not a crustacean contrary to what its name suggests, and it does not breathe thanks to the reversible binding of oxygen to iron. Its respiratory pigment is not haemoglobin but haemocyanin, consisting of two Cu atoms bridged by the dioxygen molecule with each of the Cu atoms directly linked to the protein chain by three nitrogen atoms from histidine residues. The term cyanin refers to the blue colour of the complex and has nothing to do with the cyanide ion met earlier in this book. The haemocyanin molecule is contained in mobile cells known as amebocytes (they swim around like amoeba) that circulate in the horseshoe crab's blood. When O_2 is attached to the haemocyanin, the Cu atoms are in the 2+ oxidation state and appear blue, while without O_2 the blood is colourless and contains Cu atoms in the 1+ state (Figure 9.3). The respiratory function of haemocyanin is similar to that of haemoglobin, but it is only about one-fourth as efficient as haemoglobin at transporting oxygen. This explains why octopuses (which also breathe with haemocyanin) are quiet and rather static animals that get tired easily and quickly run out of breath if you bother them during a dive.

But let's go back to the horseshoe crab blood as it has a fantastic property, although not correlated to its metal content. Horseshoe crabs do not have an immune system as we have, protecting them from infections. Instead, they prevent the entry of infected particles through wounds by coagulating

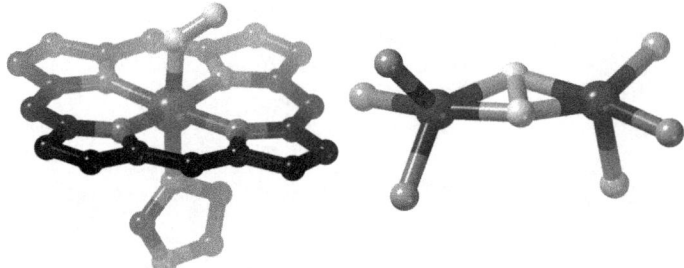

Figure 9.3   Note: hydrogen atoms have been left out of these models. Left: the heme group charged with a dioxygen molecule. It is attached to the big haemoglobin molecule by the group pointing downwards.[¶] Right: The haemocyanin group that blue-blooded creatures such as the horseshoe crab use to capture dioxygen. Two copper ions hold the O_2 molecule sideways, completely differently to how it is held in haemoglobin. The atoms to the left and right are the nitrogen atoms of the histidine groups (not shown) attaching the copper atoms to the protein (just as in heme).

amebocytes locally, hence defending the organism against pathogens. This property is used in medical applications to detect bacterial endotoxins and there is a huge demand for a derivative of the horseshoe crab blood called *Limulus* amebocyte lysate (LAL), to test medicines and chirurgical material for bacterial contamination.

This has resulted in the capture of more than half a million crabs annually to produce LAL for the pharmaceutical industry in the US. They are collected and are bled of their precious blue blood before being released back to the sea. The mortality rate for the horseshoe crabs is about 30%, but the process generates a considerable profit for the industry, as LAL is valued at around US$ 15 000 per litre and each crab can give about 200 millilitres of blood per bleeding session.

Another haemocyanin money-making industry is that of KLH, the acronym for Keyhole Limpet Haemocyanin. The keyhole

---

¶The structure of the entire molecule with four protein subunits and four heme groups was determined by a team led by Austrian born Max Perutz at Cambridge and gave him a shared 1962 Nobel prize in chemistry, the same year as the Nobel Prize for Physiology or Medicine went to the gentlemen(?) of the DNA structure. There were a couple of ladies on Perutz team, and Hilary Muirhead went on to become a professor at Bristol University, continuing the work on protein crystallography.

limpet (*Megathura crenulata*) is a giant mollusc living only off the west coast of California, approximately from Monterey (US) to Baja California (Mexico), that breathes with haemocyanin and copper atoms. However, in this case, the haemocyanin is contained in a sugar-coated giant protein, giving it immuno-stimulating properties used for the development of therapeutic vaccines against many types of cancers. The limpets, either collected from the sea or being farmed, are bled for their aristocratic blue blood, just as the horseshoe crabs. One can obtain about one gram of pure KLH per year and per animal, with a value of more than US$ 200 000 per gram, so the going is good for these vampire companies of modern times.

9.3 THE FATE OF THE OXYGEN

For heme or haemocyanin, no matter which, grabbing an oxygen molecule is easy, it is afterwards that it gets complicated. Just as you need oxygen to make the logs burn in your cosy fireplace, or coal in the power plant, we need oxygen to "burn" sugar, fats or proteins in our body. To make a long story short here, the essence of what is going on is that we break the dioxygen molecule in two, and add two electrons to each oxygen atom, giving them the oxidation state-II. At the same time, we need to chop up all bonds to carbon in whatever source molecules we are currently deriving our energy from. This transforms carbon from the lower to higher oxidation states, finishing when we have a carbon combined with two oxygens forming carbon dioxide, CO_2. Taking oxygen as -2, this will give the $+4$ oxidation state to carbon, and all valence electrons are gone (see further in Appendix 1.)

This requires several steps, and as we cannot have loose electrons on the run in our cells, we also need a secure system for transferring them from molecule to molecule. Some metal ions are ideal for this, as they can easily change oxidation states without gaining or losing too much energy. Therefore, we typically find iron going from $+2$ to $+3$, and copper going from $+1$ to $+2$ in the complicated system we call the electron transport chain, with enzymes known as the cytochrome c family.

This system sits in the membrane walls of the mitochondria, and one thing they do is to pump H^+ ions out from the mitochondria and into the surrounding cell. This is a bit like pumping water up to a power station dam, because in another part of the membrane separating the cell and the mitochondria sits an enzyme that allows them to fall back in again and level the concentration. Just like with a waterfall this can be used to produce something, and in this case, it is the universal energy currency of the cells, the ATP molecules, that come out.$^{\parallel}$

9.4   BEETLEMANIA

Not only is the oxygen processing cumbersome and difficult, moving so many electrons and breaking so many chemical bonds, it is also highly dangerous. If any of the very reactive molecules that are produced when dioxygen goes stepwise from O_2 to H_2O go astray in the body, they might wreak havoc on other sensitive systems. Hydrogen peroxide might be formed, H_2O_2, or the famous "free radical", that is the neutral OH molecule, (OH^- that has lost an electron and now has an odd, unpaired electron or dioxygen having gained one electron) or the superoxide anion O_2^-, also a radical, might appear. We, and all living things dependent on oxygen for living, need protection from these dangerous oxygen molecules.

One of the ways our bodies handle this is by using the catalase enzyme, the rogue-oxygen watchdog.** It contains four iron-heme units, closely related to those in haemoglobin, and very efficiently catalyses the breakdown of hydrogen peroxide to water and dioxygen, preventing it from splitting into two dangerous OH-radicals. Especially high concentrations are found in the liver.

The effect of this enzyme is visible with the naked eye because it is what prevents our hair from going grey. Hydrogen peroxide that we produce naturally is constantly gnawing at the pigments in our hair, but healthy doses of the catalase enzyme breaks

---

$^{\parallel}$This is one of the discoveries of Yellapragada Subbarow whom we met in Chapter 9. The reaction involves adding a PO_4^{3-} group, a phosphate ion, to a pair of such ions already attached to a DNA base and a sugar, adenine and ribose, with diphosphate, $-PO_2-O-PO_3$ dangling end. Adding another phosphate group to give $- PO_2-O-PO_2-O-PO_3$ requires energy that can later easily be released by reversing the reaction. In a further completely irrelevant coincidence, this molecular machine stranding the membrane (we call it ATP synthase) actually has moving and rotating parts just like a water-wheel or a turbine.

**The other important example is superoxide dismutase that deals with the O_2^- radicals. In humans, these contain either copper and zinc, or two manganese ions.

it down before it does too much damage. Artificial dosing is another matter, and the poor enzyme must throw in the towel and admit defeat against the platinum blonde.

The bombardier beetle on the other hand does no such thing, but instead uses hydrogen peroxide and the catalase enzyme to its spectacular advantage. In two separate compartments it stores hydrogen peroxide and hydroquinones in one, and catalase plus another metal dependent enzyme called peroxidase in the other. It is like storing the rocket fuel liquid hydrogen and liquid oxygen in the same tank of a rocket without them reacting and setting everything ablaze. This works because on its own, hydrogen peroxide is stable. Add catalase and it will produce heat and dioxygen, the dioxygen will oxidise the hydroquinones giving more hydrogen peroxide, and the catalase produces even more gas and heat, and the little bug sprays the 100 °C mixture over an aggressor in short bursts, of up to 700 pulses per second.

9.5 CHOPPING UP THE SPAGHETTI

It is, or so they tell us, a sacrilege in Italy to cut your spaghetti into smaller pieces before eating it, but that is not how our body sees things when it comes to proteins. These huge molecules are composed of amino acids, snapped together by the protein factory enzyme called ribosome. In one end, carrier molecules come with amino acid after amino acid, inside they are added to one another, and at the other end the chain of a new protein is extruded as it grows longer, normally to the size of several hundred amino acids.

The "snapping together" works by coupling the ends of the amino acids together. Each has one end with a carboxylate, $-COO^-$, and the other with an aminium ion, $-NH_3^+$, and they come together forming a peptide bond, $-CONH-$, also producing one water molecule for each bond.[††] Energy-wise this is an uphill

---

[††]This is what we call a condensation reaction. Possibly you have learnt that amino acids have one acidic end, $-COOH$, and one amine end, $-NH_2$, but that is only very formally and used to be the most common error in biochemistry textbooks. Carboxylic acids are fairly nasty compounds (*i.e.* concentrated acetic acid, CH_3COOH, a liquid), and amines even worse (such as ammonia, NH_3, a corrosive gas). Amino acids on the other hand are nice, mostly white, crystalline powders, and normally easy to handle because the acid, $-COOH$, has reacted with the base, $-NH_2$, forming a carboxylate ion and an ammonium ion, much like a salt, but the charges are at different ends of the same molecule.

process, so the body pays one ATP molecule for each peptide bond that is formed. But this is done already in the preceding step when the amino acid is attached to an RNA carrier molecule.

However, we also need to break down the proteins we eat, we cannot use them as they are. So, there is no way to respect Italian etiquette, we will have to chop them up into smaller parts. It really should not be that difficult, because if adding them together is uphill, taking them apart will be downhill, spontaneous even, and you wonder if the 18th century aristocrats were on to something when they didn't take any baths. Thermodynamics tells us our proteins should just dissolve, and that cannot be very healthy.

In real life, this process is exceedingly slow, so we need not worry. It is even so slow that the body needs special enzymes to do the chopping up. These are known as proteases,[‡‡] and are used in many bodily processes, not just digesting food. A large tribe of these are the metalloproteases, and many of those need zinc 2+ ions to work, but a few use cobalt 2+ instead. The way they do this is by gearing up the water molecule that we saw leaving when we were making the peptide bond. We now need it to come crashing in to break the bond instead.

In a beaker, we could do this by adding a very strong base, potassium hydroxide for example (caustic soda), so that instead of water we'd have hydroxide ions attacking the peptide bond. In our body, such a reagent would cause major damage, so nature has invented another way. The water molecule will hook up to a Zn^{2+} ion in close proximity to the bond to be broken, and nearby amino acid residues with negative ends will attract the hydrogens of the water molecule (Figure 9.4). It's a bit like the water oxygen atom getting partly stripped of its clothes, exposing its bare and reactive electrons.

At the same time, the protein to be cut up will be attracted to the positively charged metal ion. This serves both the purpose of positioning the peptide bond correctly for the attack from the electrons of the half-nude oxygen atom, and weakening the carbon–oxygen double bond, the first step in detaching the two amino acid residues from each other.

[‡‡]Also called peptidases or proteinases, and in the stomach we have the subclass of pepsins.

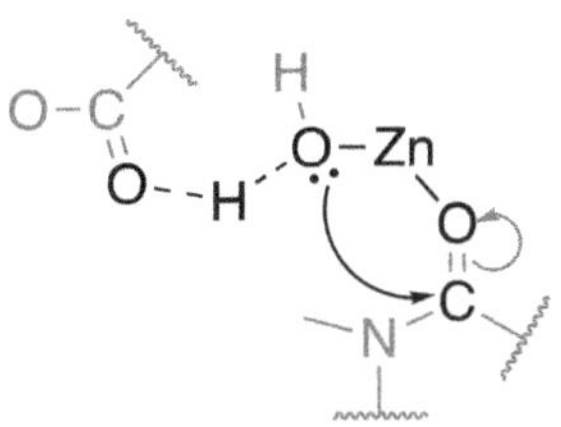

Figure 9.4 A water molecule bound to a Zn^{2+} ion becomes activated by hydrogen bonding to an organic part of the enzyme and can now act like a hydroxyl ion, its electron pair (the dots) "attacking" the carbon of a peptide bond thereby starting a chain reaction of electron-pair moves (curved arrows) that eventually leads to chopping-up of the protein. Wobbly lines indicate that only a part of the molecule is drawn.

9.6 THE GREAT BUFFALO HUNT

Strange habits are often not very strange once the reasons have been explained. But western colonialists have often payed dearly for not listening. The missionary Nathan Price does so in *The Poisonwood Bible* (Barbara Kingsolver, 1998), spectacularly failing his demonstration garden in the Congo. But for Lt. John J. Dunbar, the great buffalo hunt in *Dances with Wolves* (Kevin Costner, 1990) ending with a feast of raw buffalo liver only perplexes him. Eventually, the hunt becomes a turning point in Dunbar's relation with the Lakota Sioux, he becomes one of them and they give him the name eponymous with the movie. Hopefully, this also meant raw liver was added to his diet, especially after hard winters when food was scarce, because it was an essential source of iron, vitamin A and vitamin B12.[§§] Nowadays you do not need to hunt buffalos to get your vitamin B12 fix, vitamin B12 supplements are something you can buy in any pharmacy, and if you are a vegan this is probably a good idea, as non-animal sources are even more scarce than buffalos on the Great Plains today.[¶¶]

Cows, sheep, and other ruminants, including buffaloes, make vitamin B12, also known as cobalamin, provided they get enough of the transition metal cobalt in their food. Or rather, they don't,

---

[§§]We do, however, recommend it to be cooked, whether it is buffalo or not.

[¶¶]The buffalo hunt in Dances with Wolves takes place around 20 years before "The last buffalo hunt" when hunters from the Lakota and Yanktonais people killed around 5000 buffalo. This and similar hunts in 1882–1883 were effectively the end of the raw liver feasts and a way of life.

what happens is that they absorb this compound as it is produced by a colony of helpful bacteria they carry in their guts. Bacteria, and some other single cell organisms, are the only living creatures that can produce this complicated molecule.

The nations of the Great Plains did not know about vitamin B12, but they knew what was good for them, and the rest of the world were also slowly finding out. Pernicious anaemia was a very serious disease identified early in the 19th century, usually killing its victims within a few years. Now we know it as one of the major causes of vitamin B12 deficiency, and the discovery that you could use first raw liver, and later liver extracts, to cure the condition was awarded the Nobel prize in medicine in 1934.[III]

But raw liver was not to everyone's taste, and the liver extracts were cumbersome. Could one perhaps instead find, and use, the liver molecule responsible for the cure? This is of course around the same time that folic acid is discovered. The liver needed to be separated into its molecular components, a tedious but fairly straightforward task. The problem was to identify which of these fractions contained the active component. Feeding the different fractions to patients and watching who got better and who didn't, was at best very slow and at worst highly unethical.

It is of course a complete coincidence that the lady who came up with the solution originated from North Dakota, not that far from the Standing Rock Indian Reservation where in 2016 the latest, but surely not the last, confrontation between the First Nations, and US State and Federal government took place. Mary Shaw Shorb (1907–1990), however, was born some 20 years after the buffalo had disappeared and she was instrumental in uncovering the secretes of cobalamin.

Working at the University of Maryland, Shaw discovered that the *Lactobacillus lactis* Dorner bacteria needed the same thing from the liver as us humans. To astonished colleagues, she could in 1946 correctly identify active and none-active liver extracts

---

[III]To US scientists George Hoyt Whipple, George Richards Minot and William Parry Murphy. However, George Whipple worked closely with Frieda Robscheit-Robbins (1893–1973) in "...one of those great creative partnerships in medicine," (*British Medical Journal*, G H Whipple obituary, 21 Feb. 1976) so whether the credit was shared correctly in 1934 probably remains an open question.

within a few hours, just by observing how her bacteria grew, or did not grow. From there on, the road was laborious but fairly well mapped out. Various chemical methods were used to separate the liver components in finer and finer fractions with less components, until one day, two years later, red crystals appeared, and the *anti-pernicious anaemia factor* now called vitamin B12, was found.

But what was it? Several research groups on both sides of the Atlantic had a go at this, but it was Dorothy Crowfoot Hodgkin*** and her group in Oxford that pinned it down. In the beginning of the 1950's, a frustrated Australian named Jack Cannon had tried to grow good crystals of a molecule that had been chopped off and modified from vitamin B12, but still contained the cobalt part, but to no avail. Good crystals were a prerogative for finding out the structure and exact chemical composition of such a complicated molecule as vitamin B12, but nothing had worked for him. Finally, he just went around the lab, collected whatever different solvents he could find and threw the compound in and let it dissolve. He then left Oxford for a, as we shall see, well earnt holiday in continental Europe.

Upon his return, beautiful red crystals were waiting for him in one of the containers. His professor, Alexander Todd, immediately sent them over to Hodgkin and she gave them to her student Jenny Pickworth (later Glusker) as her research assignment. Pickworth, and the rest of Hodgkin's group, went on to "solve the structure" as we say, using X-ray diffraction, an extremely laborious task at a time when computers were only in their infancy. After un-picking what the central part of the vitamin looked like, this knowledge was applied to the X-ray diffraction data from a crystal containing the intact vitamin B12, and out came a beautiful molecule (Figure 9.5).

Knowing the structure and formula made large scale manufacturing possible, and today we make around 10 tons a year of this life-saving substance. But the path to fame for Hodgkin and her team was not a yellow brick road, as some, notably their Oxford

---

***She was awarded the 1964 Nobel Prize in Chemistry for the use of X-ray crystallography to determine the structures, and thus the chemical identity, of some of the most complex molecules known at the time: penicillin, vitamin B12 and insulin.

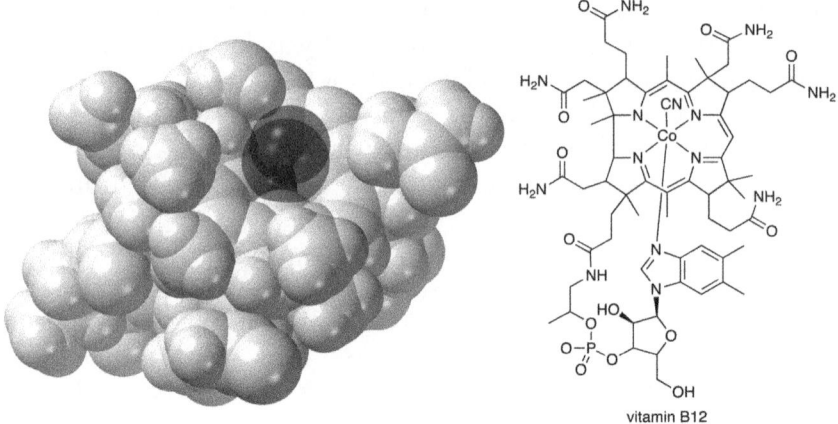

vitamin B12

Figure 9.5   Structure of the cyanide version of vitamin B12 or cyanocobalamin. Hidden deep in the centre is a cobalt +3 ion (black) binding the cyanide (black transparent). The core of the surrounding ligand (white) looks a bit like, but is in fact distinctively different from, the porphyrin in haemoglobin.

collaborator biochemist Alexander Todd,[†††] by then already "Sir Alexander", were not too keen to acknowledge the hard, and absolutely crucial, work of the crystallography group. Before their joint set of articles were published in 1955, Hodgkin got wind of a prior and un-agreed announcement Todd was to make of the formula at a meeting in Exeter. She rushed down to the south of England and made some, one supposes, rather harsh, clarifications on who had done what from the back of the lecture hall. Apparently, Todd was not embarrassed, so Hodgkin had to stay vigilant. In the words of Jenny Pickworth Glusker: "...that summer, Dorothy made sure that whenever anyone from Todd's lab gave a talk in Europe someone had to attend, stand up at the end of the lecture and clarify how the chemical formula was found."[1]

In the body, vitamin B12 is used for making molecules. It is needed as a plug-in for several enzymes, what chemist call a co-factor, where it ensures, among other things, DNA synthesis and regulation, nervous system function, and the formation of red blood cells. In our body the cyanide ion of the vitamin B12 supplement, or the versions of the molecule indigested with buffalo liver, will be exchanged giving various slightly different co-factors with many functions.

[†††]When he was awarded the Nobel Prize for Chemistry in 1957 "for his work on nucleotides and nucleotide co-enzymes", this was a different story. No quarrels there.

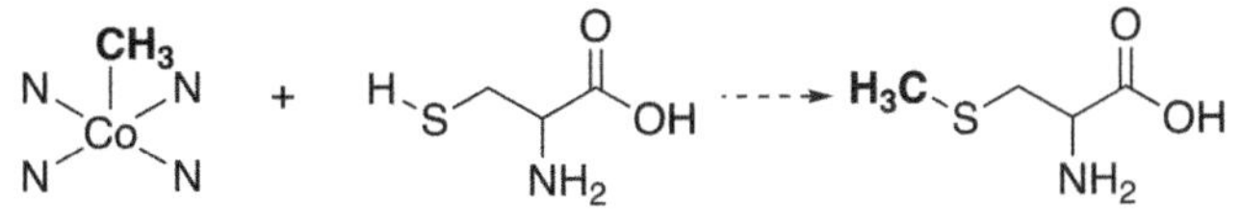

Figure 9.6 An example of a reaction catalysed by vitamin B12 (schematically shown only with four nitrogen atoms, the cobalt ion, and the – CH_3 group) together with the enzyme methionine synthase. One amino acid is transformed to another by the move of a methyl group (–CH_3) from cobalt to sulfur.

One characteristic reaction relies on the formation of a cobalt–carbon bond when the cyanide ion has been kicked out by a –CH_3 1– anion (we call this a methyl group, a carbon with three C–H bonds and an additional electron pair sitting on the carbon). The cobalt atom can now easily move this methyl group to an organic molecule, something that is tricky to do with molecules only containing C, H, N and O. For example, in our body the essential amino acid cysteine, ending with a HS– group (a thiol) can be transformed to another essential amino acid, methionine, ending with a S–CH_3 group (Figure 9.6).

This is a tricky reaction to perform, and if you remember back in Chapter 6, when we were making chewing gum, we used a methyl group attached to a lithium ion. A very efficient reagent indeed, and somewhat dangerous, although not as bad as some if its cousins (*tert*-butyl lithium is very pyrophoric and burns violently in contact with air). So, no wonder the body must use a very sophisticated machinery to do the same thing under much milder conditions without the risk of internal combustion. And while reagents and catalysts with metal–carbon bonds have become increasingly used in industry and the research laboratory, vitamin B12 is still a rare example in biology.

9.7 AT THE END OF THE STORY

In 1754, the young[‡‡‡] Swede Bengt Andersson Quist makes an important discovery; the minerals molybdena and plumbago are not the same thing, and neither contains lead as was supposed previously. Plumbago is what we now call graphite, but the molybdena

---

[‡‡‡]Although not as young as some sources make him out to be. Bengt Andersson Quist was born in 1726, not 1729, and the work on the two minerals was published in 1754 not 1745. Svenskt biografiskt lexikon, Swedish National Archives, https://sok.riksarkivet.se/Sbl/Presentation.aspx?id=7466, retrieved 4th Feb. 2018.

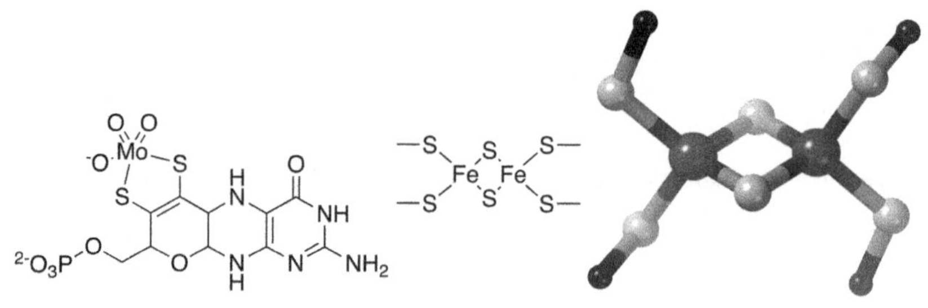

Figure 9.7   The molybdenum cofactor (left) and the iron-sulfur compound (middle and right, the carbon groups at the ends are shown schematically only) in the enzyme that oxidises xanthine to water soluble uric acid that can be removed from the body in the urine.

contained sulphur and a metal that was neither iron, copper or tin. The oxide of this metal was produced in pure form by another Swede, Carl Wilhelm Scheele, and he named it molybdenum (or rather the Swedish and German equivalents molybden and Molybdän).

We need such tiny amounts of this metal that it is almost impossible for humans to have a molybdenum (Mo) deficiency, except perhaps after being on a drip-feed for a long time. But nevertheless, it performs some very important reactions in our body. We started this chapter with breathing oxygen, used this to burn sugar to get energy, ate proteins and had them chopped up into smaller pieces and then used vitamin B12 to make bigger molecules again, and we will end with one molybdenum function that gets rid of unneeded components. This molecule might cause big trouble on its way out of the body, and we have already met it before, it is the urate ion, the anion of uric acid, and in Chapter 2 we learnt that it could cause kidney stones.

Molybdenum plays a key role in uric acid formation, and without it, you would probably have bigger problems than kidney stones. Xanthine oxidase is the enzyme that performs this trick, xanthine being the molecule that is oxidised to uric acid, and the enzyme needs a molybdenum loaded co-factor as well as a funny looking iron-sulfur compound, both of which are displayed in Figure 9.7.

9.8   THE PERIODIC TABLE OF LIFE

At the end of this chapter we should emphasise what we said at the beginning: This is but a small sample of the work metal ions do for us in our bodies, and our knowledge in this area is

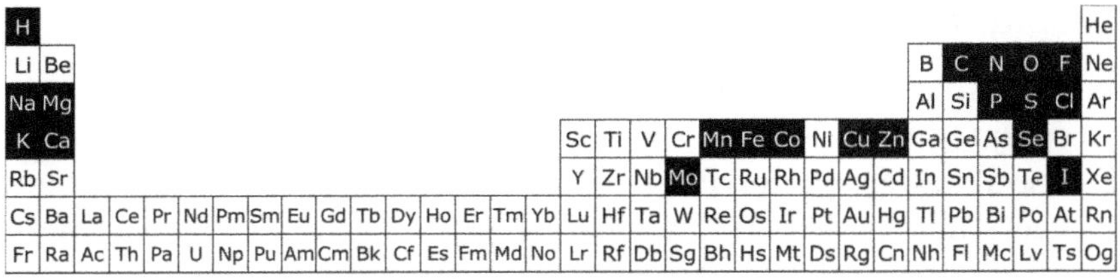

Figure 9.8   White-on-black, essential elements in humans as we know today.[2] The entire Periodic Table of Life is, however, a bit larger as other biological organisms may require one or more additional elements for some selected functions, even as far "down" as the lanthanides.

still incomplete. It is not just the question of what, and how, they do this when they are in their proper places in enzymes and co-factors, it is also the question of how they get there, how their concentrations are regulated and how, and if, they are stored somewhere. Figure 9.8 is a periodic table in which we have indicated, as far as we know today, all elements with a necessary function in our bodies.

Some of these metal ions are excellent construction materials, calcium ions such as in calcium phosphate for our bones, and calcium carbonate in limestone and eggshells. Iron is popular with humans, especially in combination with carbon to form steel, but is not used in our own biological systems in any materials capacity. In mussels, however, the situation is different and iron-catecholate type compounds (as shown in Figure 5.9 and the anthrax siderophores) are used as efficient cross-linkers when the mollusc attaches itself to a rock or other surface. This kind of glue can also be used as an inspiration to modern materials chemists,[§§§] but in the next chapter we will turn first to another type of material, paint, and we will once again take to the sea.

[§§§]*Mytilus californianus* and *Mytilus galloprovincialis* (the Mediterranean mussel) have threads containing considerable bonding between the abundant catecholic amino acid 3,4-dihydroxyphenylalanine (also known as DOPA, the starting material for several neurotransmitters used in our body, here in a completely different role) and iron(III). It was put to use in the field of materials chemistry by Emmanouela Filippidi, Megan T. Valentine and co-workers: *Toughening elastomers using mussel-inspired iron-catechol complexes*, E. Filippidi, T. R. Cristiani, C. D. Eisenbach, J. H. Waite, J. N. Israelachvili, B. Kollbe Ahn and M. T. Valentine, *Science*, 2017 Oct 27, **358**(6362), 502–505.

REFERENCES

1. J. P. Glusker, *ACA History*, 2011, http://www.amercrystalassn.
   org/jp_glusker.
2. Food Labeling: Revision of the Nutrition and Supplement
   Facts Labels, US Food and Drug Administration, Department
   of Health and Human Services, 2016, https://www.gpo.gov/
   fdsys/pkg/FR-2016-05-27/pdf/2016-11867.pdf.

CHAPTER 10

Vanishing Warships and Sugary Gold

Noël Coward, the famous English playwright, composer, director, actor and singer, narrowly lost out on the Best Picture class of the 1943 Academy Awards. The "Oscar" instead went to *Casablanca*, and few people would argue with that decision. Coward directed the film *In Which We Serve* that was likewise nominated for Best Original Screenplay (also by Coward and no award either) and he even played a lead character, Captain Kinross.[†] In some respects, however, the real protagonist of the film is the destroyer HMS *Torrin*, but as the movie is in black-and-white, we can't really tell its colour.

10.1 UNCLE DICKIE'S PINK BOATS

Colour is perhaps the most tangible part of chemistry in materials science, and in this chapter, we will first encounter metal ions in pigments, the colour-giving part of a paint. The question of the colour of HMS *Torrin* arises, as outside of the movie narrative, Captain Kinross, in the real-life version of Louis Mountbatten,

---

[†]Sir Noël Peirce Coward (1899–1973) did, however, get an Academy Honorary Award for his naval drama.

The Rhubarb Connection and Other Revelations: The Everyday World of Metal Ions
By Lars Öhrström and Jacques Covès
© Lars Öhrström and Jacques Covès 2019
Published by the Royal Society of Chemistry, www.rsc.org

had already made his very own visible mark on the Royal Navy, and somewhat surprisingly perhaps, it was pink.

A great many things have been said about Louis Mountbatten, and not all of them nice by far. In the historical drama television series *The Crown*, he is a shrewd behind-the-scene fixer, colloquially referred to by Queen Elizabeth and her closest as "Uncle Dickie". His main achievement appears to be getting his nephew married to the queen-to-be, but to his credit one should probably point out that contrary to most members of the leisure class, he had a real job for his entire life. He started out in the Royal Navy at the age of 13 and retired as Chief of the Defence Staff.

In 1940, possibly while commanding the HSM *Kelly* (the real HMS *Torrin*), he observed that a pink merchant ship his destroyer flotilla was escorting almost disappeared in the dim light of dusk or dawn. Interested as he was in innovations and technical development (he was a member of the Institution of Electrical Engineers), he had a suitable colour mix prepared and his entire flotilla were then painted in Mountbatten Pink, a grey paint with some Venetian red added, giving it a greyish-mauve colour, perfect for blending in with a pink sunset.[‡] And the pigment Venetian red, regardless of the fancy name, is nothing else but simple iron oxide, Fe_2O_3, in its most common magnetite form. This material has iron 3+ ions surrounded by bridging oxygen 2− ions in a three-dimensional network, and you know it as one of the components of common rust.

However, there are ways to see you even in the dark or if you are pink. Radar was one of the many technological advances coming of age just before the Second World War. It is based on the ability of radio- or micro-waves to bounce back from solid objects and being detected by a receiver system. Obviously, as soon as this became widespread, military boffins started to think about ways to paint a ship or an airplane to make it invisible to a radar system.

Modern so-called stealth technology is based on many different components, not just paint. Odd shapes are one feature, as used in the experimental US vessel *Sea Shadow*, the inspiration for the evil villain's stealth ship in *Tomorrow Never Dies*, Pierce Brosnan's second James Bond movie. The Swedish Royal Navy's corvette HMS *Visby*, one of the world's first commissioned stealth

[‡]Colour code #997A8D.

ships, has a non-metallic hull and avoids right angles. The specifications, as presented by shipbuilder Kockums,[1] do not include any mention of stealth-paint, but probably the pamphlet does not tell us everything. What we wonder, for the purpose of this chapter, is if HMS *Visby* is in critical places coated with the oddly named material known as iron ball paint.

Iron ball paint contains very small dispersed particles of iron, whose magnetic properties will make them adsorb the radio- or microwave radiation from the radar transmitter and turn it into heat, instead of bouncing the waves back to be detected by the enemy. In order to prepare this material, very pure iron is needed, and it is here that the chemistry with metals and ligands come in. And with some quite unexpected chemistry as well, because who thought you could dissolve a metal in a gas?

But that is really what happens if you react metallic iron with the very poisonous molecule carbon monoxide, CO.[§] You then get a yellow-orange liquid composed of iron pentacarbonyl molecules, jargon for an iron atom in oxidation state zero, just as the metal, binding to five molecules of carbon monoxide, with formula $[Fe(CO)_5]$ and equally toxic. Confusingly, these types of compounds (we briefly encountered someone being poisoned by $[Ni(CO)_4]$ in Chapter 4) are known as a metal carbonyls[¶] and they are all highly symmetric molecules with only weak forces in-between the molecules, which means there is not much that will stop them from escaping and move up into the gas phase. So, iron pentacarbonyl boils at 103 °C and can be distilled, just like ethanol, which means it is easy to obtain very pure iron in this way (Figure 10.1).

After distillation, iron pentacarbonyl can easily be decomposed to metallic iron as very fine particles[‖] and carbon monoxide. This iron is, again confusingly, called carbonyl iron, but there are no CO molecules left, the name only refers to the production method. The radar invisibility paint is but a fringe use, these fine particles have many other uses, from making electronic components to iron fortification of bread.

[§]But you would not actually use the gas as such for industrial production, but pressurise it so that it becomes a liquid. The conditions needed are quite harsh, 5–30 MPa (roughly 50–300 times the normal air pressure), and temperatures of 150–200 °C.

[¶]Should be distinguished from the carbonyl group, a carbon doubly bonded to an oxygen with two additional bonds to other atoms, as in acetone for example: $(CH_3)_2C=O$.

[‖]Spherical particles of 1–10 μm in diameter.

Figure 10.1   Left: Iron pentacarbonyl boils at 103 °C and can be distilled, just like ethanol. Before General Motors started to pollute our big cities with tetraethyl lead in the 1920s, it was used as an anti-knocking agent for gasoline engines in parts of Europe. It is also the precursor for carbonyl iron used in radar adsorbing iron ball paint used for stealth airplanes and ships, possibly also on the Swedish Navy's HSM Visby shown to the right. Reproduced from https://sv.wikipedia.org/wiki/HMS_Visby_(K31)#/media/File:K31_HSwMS_Visby_(8643086211).jpg under the terms of the CC BY 2.0 license, https://creativecommons.org/licenses/by/2.0/.

The particles can also be oxidised to give modern versions of the iron oxide based pigments. A more elaborate network compound is found in the green pigment malachite, in which carbonate ions (that we also find with an added H^+ ion as HCO_3^- in mineral water) bridge copper 2+ ions (Figure 10.2).

Just like Venetian red, malachite is an old pigment and was also used on board another ship doing a vanishing act. On the 10th August 1628, the pride of the Swedish king Gustav II Adolph, the newly commissioned warship Vasa, set sail on its maiden voyage. She was one of the grandest and most heavily armed vessels of its time, but all her cannons amounted to little against the rather moderate puff of wind that sent her to the bottom of the Baltic, right in the Stockholm harbour, after only little more than one kilometre under sail.

An embarrassing catastrophe for the budding European great power, the ship was perhaps still a shrewd investment, as today she is a unique tourist attraction in central Stockholm. She was raised from her oblivion, covered by slime and 300 years of unnameable pollution from the capital, in 1961, and made a short voyage back to where she was born with the new museum later constructed around her.

Figure 10.2 Left: The structure of malachite with the detail of the carbonate bridge (C: black; O: light grey; Cu: dark grey) and how it bridges the copper 2+ ions. Centre: The network structure. Right: List of pigments procured by the navy yard on Skeppsholmen, Stockholm, in 1615. Malachite is the last entry from the top "Bärg grönt - 4 pund". Document kept at the Riksarkivet, Marieberg, Stockholm: Skeppgårdshandlingar 1615 65-4.

There is a lot of materials chemistry related to this old ship, some nicely highlighted in the exhibitions, and her colours have been a special concern. An old box cover for a 1970's plastic model of the ship shows the stern brightly painted in blue, gold and yellow, but this patriotic colour scheme has since gone through a major revision.

The wood raised from the heavily polluted Stockholm harbour had very few visible signs of paint but for bit of gold here and there on a few decorative items, the rest was murky brown oak. Microscopic paint and pigment residues were found, but methods to analyse them were yet to be developed, so samples were collected for later analysis, and then the ship was showered in polyethylene glycol (PEG) for almost two decades.**

The presumed yellow and blue colour scheme was discredited in the 1990s when samples from the ship were subjected to new analysis, and riche hues of green, red and pink were found, among them the green malachite pigment. But it was not only hard core science that effectuated this change in thinking. Even though Vasa is the only surviving ship of its kind, the art and the decorations were not created in an artistic vacuum. Comparison with contemporary sacred art and palace decorations, as well as 17th century maritime paintings of similar Dutch ships, also made the Vasa historians and scientists rethink the colour scheme. In addition, bureaucracy helped, as can be seen in a list of pigments acquired by the navy yard on Skeppsholmen (right opposite the Royal Palace in Stockholm).

The way we understand the significance of the decorations has also shifted. From a more warlike interpretation to the modern view of the purpose as a grandiose glorification of the king himself. Modern navies keep their ships in port as much as they can because they are expensive to run, and this was also the case in the 17th century. This made the powerful story told on the high-rising stern of the might and glory of His Majesty King Gustavus Adolphus, "The Lion of the North", a prime propaganda weapon to be displayed in the major ports of the newly minted Swedish empire.

**To keep the wood from cracking and disintegrating as it dried. No similar conservation efforts on this scale had ever been undertaken before, and lessons learned by the Vasa museum conservators have greatly profited subsequent projects such as the raising and conservation of the English warship the *Mary Rose*, pride of Henry VIII, that sank in 1545 and is now located at the Historic Dockyards in Portsmouth, UK. See also the official museum site: https://www.vasamuseet.se/en.

However, the high-ranking officers on board seemed to need no reminding of their supreme commander's abilities, however. Despite extensive search, the Vasa scientists have found no traces of any pigments inside the cabins and have concluded that the interiors were just plain wood with varnish. But it is also a fair guess that officers spent as little time on board as possible.

Henrik Hybertsson, the Vasa's master shipwright, inconveniently died before the ship was completed, leaving the management of the yard to his widow, Margareta Nilsdotter. This may seem unusual for its time, but it appears women often occupied more prominent roles in many parts of society than later history has granted them. No doubt Mrs Nilsdotter had already taken a keen interest in her husband's affairs, even before the long illness that preceded his death in 1627. In addition, in those turbulent times, many women of the nobility managed large estates while their husbands were away looting and pillaging in present day Germany, Poland and the Czech Republic.[††]

10.2   THE ENTREPRENEUR COUNTESS

Not that it ended well for all these husbands. Some 70 years later, Count Carl Piper, in all but name the Swedish prime minister, went out to war with the King Charles XII, leaving his much younger wife Christina Törne to run his affairs and raise a family of four small children. He was never to come back, although his wife would occasionally visit him during pauses in the campaigns. Having been taken captive at the, for Sweden, disastrous battle of Poltava in southern Ukraine 1709, he perished as a Russian prisoner of war in 1716.

The estate and his large business interest were in good hands, however, as Countess Piper became one of the most successful entrepreneurs of 18th century Sweden.[‡‡] In 1725 she made an investment that would have long lasting effects visible to this

---

[††]Alternatively fighting for the noble cause of Lutheran Protestantism backed by Cardinal Richelieu, who evidently was much more French than he was a catholic. Either way you look at it, the consequence was that Swedish churches and palaces are now filled with as many items of questionable origins as the British Museum.

[‡‡]To show that Christina Piper, as she is mostly known, was not a unique example we can also mention that in the generation preceding her, Maria Sofia De la Gardie, Countess Oxenstierna, was at the time one of the major players in the Swedish manufacturing industry and banking.

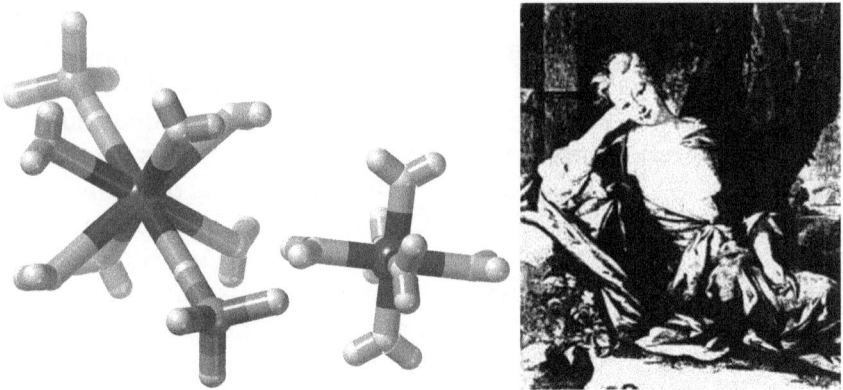

Figure 10.3   Left: Potassium aluminium sulfate dodecahydrate, also known as alum, contains larger units of $[K(H_2O)_6(SO_4)_2]^{3-}$ ions (left) and smaller $[Al(H_2O)_6]^{3+}$ ions (right). The value for countess Christina Piper, pictured right (painting by David von Krafft), lay in the aluminium ions, ubiquitous in aluminium oxides, but hard to get on their own as "free" ions.

day: she bought the faltering Andrarum works in the recently occupied province of Scania in the south of Sweden, and turned the manufacture of potassium aluminium sulfate dodecahydrate, also known as alum, or to be more precise potassium alum, into a thriving business that survived for almost two centuries (Figure 10.3).[§§]

In the bathroom, we may use this compound to stop minor bleeding from a razor cut, but curiously enough, this chemical also turns up in many "aluminium free" deodorants. These work the same way as any standard deodorant containing, for example, aluminium chloride. Ostensibly the reason is because potassium alum can be found in nature, as the mineral alunite, so therefore it is "natural", never mind that it contains exactly the same aluminium ions, $[Al(H_2O)_6]^{3+}$, as the aluminium chloride hexahydrate often used in other deodorants.[¶¶]

Presumably, it was not concerns about body sweat and odours that made alum such a big issue for such influential and sometimes dangerous people such as the Italian banking pioneers of the Medici family, the infamous Borgias, including one or two popes,

[§§]The alum story is partly based on a podcast in *Chemistry in its Element* series by the RCS's Chemistry World: https://www.chemistryworld.com/podcasts/alum/7068.article.
[¶¶]To be fair, there are also deodorants that are truly aluminium free.

and the wealthy burgers of Flanders, Netherlands and England. In the 15th century, the Medicis and the Papal States (this family also produced a few popes) even tried to establish a European monopoly through its all-penetrating network of businessmen and courtiers. The reason? Alum was the only source of soluble aluminium 3+ ions, making it an excellent mordant, a compound that helps fix a dye to a textile preventing it from being washed out by water, like the famous baby with the bath water. In medieval Europe, the wearing of some colours was often regulated in law, so to pay to stand out of the crowd was not only vanity but a social necessity for some. This made coloured cloths a big business in medieval Europe. As we say today, the dying processes made possible with aluminium 3+ ions created an enormous add-on value for the wool and cloth industry in medieval Europe.

The metal ion attaches to the pigment molecule and drags it into the fibres, making it "bite", thus the term mordant, by precipitating the ensemble as insoluble polymeric hydroxides, efficiently hindering the dye from leaching out during washing.[III]

No colour, except perhaps the blue lapis lazuli obtained from the far away mines in the Afghan mountains and therefore known as ultramarine, was more exclusive than the red or crimson you could get by boiling cochineals, a scale insect from South America, in water (Figure 10.4). In the words of Amy Butler Greenfield: "To obtain it, men sacked ships, turned spy, and courted death".[2]

One might wonder what the problem was, as aluminium is a very abundant metal, only oxygen and silicon are more common in the earth's crust. It is literally all around us. The thing is that the Al^{3+} ions are almost always tightly bound to oxygen in aluminium oxide, Al_2O_3, an insoluble and very hard material, and the mordant needs to contain free Al^{3+} ions.

The mineral alunite could be picked as rocks in the Egyptian western desert, but that was a scarce and far away resource. Although nobody at the time realised what this compound was, aluminium as a metal and an element was not to be discussed until the mid 18th century, industrialists still managed to make alum in large quantities from something called shale, a very

---

[III]Much the same mechanism, precipitation, is used when wastewater is treated with alum. Aluminium hydroxides are precipitated out together with various pollutants to give an easily separated solid phase. This, as well as tanning, were also ancient uses of the compound.

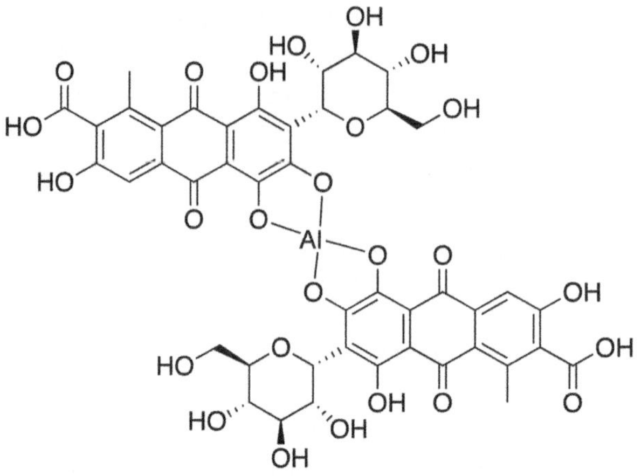

Figure 10.4   The free acid is known as carminic acid and the aluminium compound is carmine. The drawing shows one of several possibilities of how aluminium 3+ ions can bind to carminic acid. The exact structure of these compounds is not known.

variable resource of mud and clay with some organic matter embedded. The Medici family's alum cartel, forcing high prices on the cloths manufacturers in England and Flanders, thus did not last. Moreover, in England, the divorce problems and subsequent reformation of Henry VIII caused strains with many catholic states, stimulating a domestic industry based on shale, established on the Yorkshire coast at the end of the 16th century.

Making alum from shale is a more difficult process than picking alum rocks in the Egyptian desert, and the success of these alum works is remarkable, given the limited knowledge of chemistry at the time. The short version is that the aluminium- and iron-containing shale, with a substantial organic component, was burned on a wood fire. This very slowly oxidised the iron pyrite, or FeS_2, to iron 3+ ions and sulfur oxides. The soluble metal ions, mostly iron, potassium, sodium and aluminium were then extracted out with water that also transformed the sulfur oxides to sulfate ions. This liquid was concentrated and finally crystals precipitated out. Normally, potassium alum was produced, but in the case of the first Yorkshire works where urine was also used as a reagent, ammonium alum was the final product.

Christina Piper's venture in Andrarum was probably motivated by similar concerns for domestic production, and giant heaps of

Figure 10.5   As we nowadays have easy access to aluminium 3+ ions, growing alum crystals has become a popular school experiment, and
nationwide competitions are often held. Left: Jason Benedict,
chemistry professor at the University at Buffalo, runs the U.S.
Crystal Growing Competition and here holds one of the 2016
entries. Credit: Douglas Levere/University at Buffalo reproduced with permission. Right: From the 2014 Crystal Growing
Competition in Abomey-Calavi, Benin. Credit: Dr Marielle
Agbahoungbata, University of Abomey-Calavi, reproduced with
permission.

the solid residues from the slow burning shale can still be seen
on the site. Today it is deceptively rural, horses walk around
in the odd landscape just by the gorges of the Verka River, but
once it was the site of one of northern Europe's largest chemical industries, where hundreds of workers toiled to produce the
colourless but often impressively large and beautifully shaped
crystals (Figure 10.5).

In the mid 19th century, pure aluminium sulfate could be
made much more cheaply and has since replaced alum in most
applications. In Andrarum, the last crystals were grown in 1912,
but the stylish little baroque palace, Christina's Court, and the
surrounding park is now open to the public and remind us of
this unusual industrial pioneer.*** In the tradition of great
estates, majestic red deer roam the large parklands surrounding
the palace. We, however, will switch our attention once again to
something substantially larger, the American bison, colloquially
known simply as the buffalo.

***Christinehofs slott, https://www.christinehofslott.se.

10.3   BUFFALO CHEMISTRY

The near extinction of the buffaloes, or American bison as they should be properly named, had nothing to do with the people of the Great Plains. But the question is; did it have anything to do with chemistry?[3]

Even if people like the Sioux could make impressive killings, especial using horses and guns, the buffalo massacres of the Great Plains were the work of non-indigenous hunters. And this happened in just a few years. By the end of the 1870's for the southern herd, and the middle of the 1880's for the northern herd (the Pacific Railroad had effectively cut the plains in two parts), most buffalos had been turned into winter coats, shoes or transmission belts in Europe. Not to mention that increasing transport required leather for harnesses and saddlery, as well as upholstery for carriages.

But why in Europe and not in the US? In 1872, The Times in London reports that US tanners simply had not succeed in tanning the buffalo hides, but that British tanners had, and other sources indicate that the Germans had already done so by 1870. There are suggestions that a major tanning innovation in Europe around this time was responsible for this, and that it took a long time for the process to cross the Atlantic. For example, in *Buffalo Hunt: International trade and virtual extinction of the North American bison*[4] Scott Taylor at the University of Calgary shows how the hides were shipped to Europe in large numbers, but the nature of the "invention" remained elusive as there seems to have been a tendency to keep trade secretes rather than taking out patents and sharing information that way. It seems clear though, that without the competence of European tanners, the buffaloes might have fared a great deal better, at least for some time.[†††]

As the use of animal skins and leather is as ancient as the hunt itself, one could be excused for thinking this was a well-known business by the end of the 19th century, but this was not the case. The process of making leather from hides was traditionally a messy, smelly and often very low-status occupation, the tanners

---

[†††]There was also a more sinister side to the buffalo hunt. While no group of hunters had any interest in exterminating the buffalo, some US officials, including general Sherman, instead saw this as an efficient way to subdue the People of the Plains, who were dependent on the buffalo for their livelihood. See D. D. Smits, *West. Hist. Q.*, 1994, **25**, 312–338.

normally being forced to locate themselves and their trade to the outskirts of towns and settlements. Not unreasonably one might add, as the tanners frequently used putrefaction, faeces, urine, and mashed animal brains in their work.

A lot of chemistry in other words. But the old processes were never very efficient, and had to be tailored for each type of dead animal used. To make a long story short, we will jump straight to the step that makes leather out of the hide, technically speaking the proper tanning, assuming we have gotten rid of all fat and hair.

This basically leaves us with a material containing four types of proteins: collagen that makes nice fibres forming the basis of the leather, and the unwanted three: elastin that does not form fibres and is insoluble in water, albumen that is soluble in water, and keratin in any residues of hair. We can get rid of the keratin and the albumen fairly easy, but elastin is trickier, and too much left in the hide will make a bad leather. The chemistry needed here is a way of breaking down the elastin into soluble parts but leaving the collagen unharmed. In the previous chapter, we talked about "chopping up the spaghetti", and such enzymes may come in handy here.

Even though our ancestors did not know about microorganisms or enzymes, they still put them to work, explaining some of the gorier ingredients in the traditional processes. Bacteria excreted with faeces have some enzymes that will act only on elastin, leaving the nice collagen fibres behind. Soaking in a mixture of dog poo and water used to be one way of clearing out the elastin, making leather manufacture one of the oldest examples of the use of biotechnology for non-food-and-drink purposes.[‡‡‡]

Once we have a material composed of more or less pure collagen fibres, there comes the tanning proper, named after the chemistry originally used. Because the word "tanning" is somehow related both to oak and *Tannen*, the German word for fir, (yes, there seems to be some botanical confusion here). In great pits, hides and the bark of oaks or other trees were layered and soaked with water for weeks or even months, depending on the thickness of the skin. Some sources say up to half a year for buffalo hides.[5]

One reason this is very slow is because of the tanning molecules. These are water soluble, but big. As molecules in solution

---

[‡‡‡]If not the oldest. Surely leather making predates gunpowder manufacture, in which the use of soil bacteria and urine were starting to be used in the 16th century. Also, in this reaction metal ions make the enzymes tick. See for example L. Öhrström, *The Last Alchemist in Paris*, OUP, 2013.

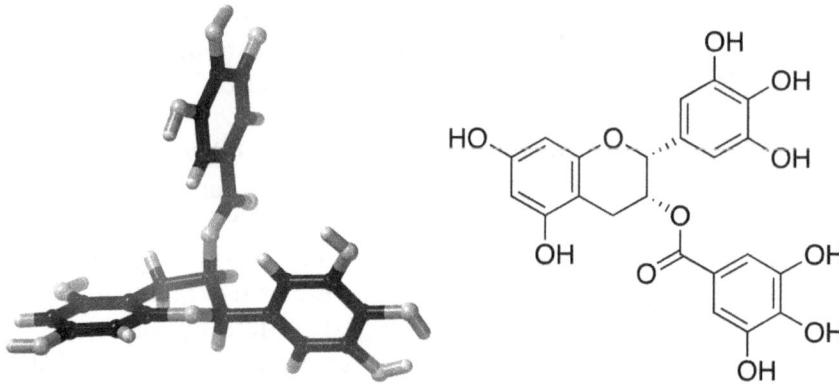

Figure 10.6   Epigallocatechin 3-gallate exemplifying a polyphenol. This molecule is found in green tea, but similar compounds from the bark of trees such as oak were traditionally used for tanning, the last step in preparing leather.

are moved around by colliding with each other in a random fashion, and a small water molecule banging into a large tannin molecule, weighing perhaps 30 times as much, will cause it to stir as much as a billiard ball hitting a bowling ball, it will take time for them to dissolve from the bark and then diffuse into the space in-between the collagen fibres.

One of the many varieties of tannin molecules is shown in Figure 10.6. They all have in common the six-membered aromatic rings (as in benzene) with many oxygens pointing outwards, mostly as –OH groups, thus they are alcohols. More specifically, –OH groups sitting on benzene rings are called phenols, and with many of those, these compounds become polyphenols.

The protein fibres have an unusual amino acid mix, containing a lot of glutamic acid residues, which is important for the tanning process.§§§ This is because the glutamic acid contains an additional carboxylic acid, not just the one that will combine with an amine to click the long protein chains together. These carboxylic acid ends are mostly turned outwards from the fibres, and will

§§§We talk about the individual amino acids in a protein as "residues", as they are no longer amino acids because the amino end has reacted with a carboxylic acid end to form an amide, known as the peptide bond. As a free molecule, glutamic acid can be made into a sodium salt, the well-known monosodiumglutamate, MSG. Naturally occurring, but also manufactured on a large scale, in both cases renowned for the umami taste it adds to various dishes and foodstuffs.

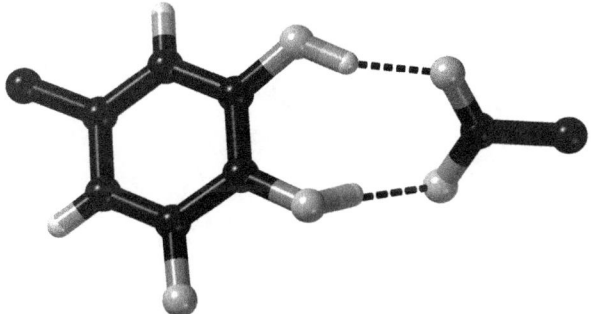

Figure 10.7   A double hydrogen bond (thus quite strong) shown as dotted lines between a tannin fragment (left) and carboxylate group sticking out from a collagen fibre (right). This is how the tannins cross-link the fibres in vegetable tanned leather. This type of hydrogen bonding between polyphenols and proteins has a lot of other implications, especial in food chemistry and happens for example when we add milk to tea, and drink red wine with red meat.

be grabbed by the added tanning substance. If these are tannins, this will mostly happen by hydrogen bonding,[¶¶¶] the moderate interaction between molecules that hold together ice for example, and the fibres in cotton and wood. And given that they are so big and have these phenol groups pointing in all directions, the tannins will bond to different fibres with different rings, we say they will cross-link the fibres (Figure 10.7). This will both protect the fibres from degradation from other substances and bacteria, and make the leather more flexible. Vegetable tanned leather will contain around 30 percent polyphenols.

But half a year is a long time to wait if you have killed thousands of buffaloes and need to turn their hides into leather. And there is one major 19th century tanning invention that greatly sped up and improved this process, and this was chromium tanning. Although, as we shall see, the timing is slightly off for blaming the near extinction of the buffalos on the chemistry of this metal.

The invention was a radically different way to solve the problem of stopping bacterial degradation of the collagen protein fibres and improving the mechanical properties of the leather by using basic chromium III sulfate, $[Cr_2(H_2O)_6(OH)_4]SO_4$, as a tanning

---

[¶¶¶]A hydrogen bond is formed between a positively charged hydrogen, itself bonded to an electronegative atom such as oxygen, and another electronegative atom. Like the dotted line here: $H-O-H\cdots OH_2$.

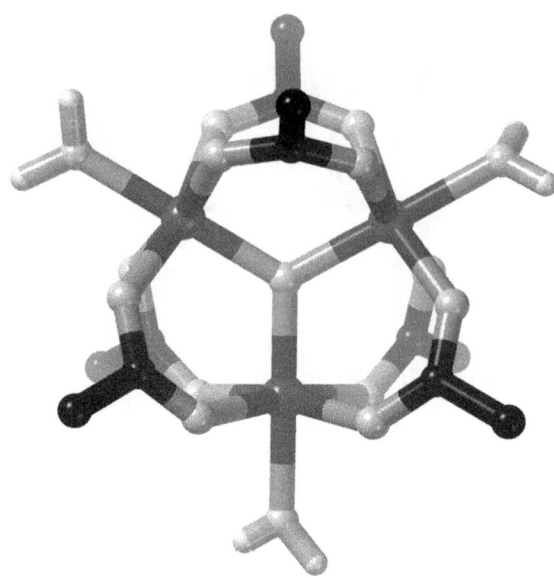

Figure 10.8 A model of a possible trinuclear chromium(III) compound bridging the collagen fibres in the leather by binding of the carboxylate ends (black) of the collagen glutamic acid residues. The atom in the centre is an oxygen 2− ion and each of the three chromium atoms additionally binds one water molecule.

agent. This salt releases Cr^{3+} ions that, even when they drag around their six coordinated water molecules as $[Cr(H_2O)_6]^{3+}$, are much smaller than the big and clumsy polyphenols, so they react faster with the collagen carboxylate groups. Several of the chromium ions team up together into dinuclear or trinuclear (Figure 10.8) units bridging the fibres, and again it is the glutamic acid residues that do the trick. Now by forming real bonds to chromium, much like the way the carbonate ions bind copper ions in the green pigment malachite. It is not only faster, it also causes less structural damage to the collagen fibres and renders the leather water resistant. The chromium content of this kind of leather is typically 2.5 percent by weight.

We suspected that the introduction of chromium tanning on a large scale made this process much simpler and faster, so suddenly great numbers of buffalo hides could be processed economically for a growing market in Europe. A neat theory, but it does not seem to hold. True, chrome tanning can be traced back as early as the 1840's (less than 50 years after chromium was established as an element), but it was a slow development, and even by 1880, when the southern herd had already been wiped out, only a dozen tanneries in Europe and one in the USA were using the process commercially. Surely not enough to have processed the millions of hides involved.

So, we can exonerate the chemists from driving the buffaloes into virtual extinction, and instead shift the blame onto the versatile and competent European tanners, used to hides from all over the world and ingenious in applying all other sorts of traditional tanning methods. On the other hand, how was a tanner in Northampton to know that the American bison that were on tour with *Buffalo Bill's Wild West* in London 1887 were some of the last of its kind?[IIIII6]

10.4 CORSET BLUES AND PERMANENT INK

The red colour of rubies comes from trace amounts of Cr^{3+} surrounded by six bridging oxide ions (O^{2-}), in an otherwise boring aluminium oxide, but chrome-tanned leather instead has a light blue colour, just as many chrome(III) carboxylates are blue. Luckily, as blue suede shoes are perhaps not to everybody's liking, the Cr^{3+} ions, just as the Al^{3+} ions, also act as a mordant, binding and fixing other pigments among the collagen fibres, so chrome-tanned leather is easy to get in any colour.

The development of chrome tanning coincides with the Victorian era, and some have even suggested that the perhaps most quintessentially Victorian garment, the corset,[7]**** had something to do with this.[3,8] The reason is that the combination of the leather used to clad the structural elements, the iron that was beginning to replace the whalebones in these, and the moist ooze coming from a tightly confined body, being salty and therefore corrosive, gave an unwanted chemical reaction.

Iron ions would be release from the corroded bracing strips, travel through the vegetable tanned leather and there combine with the tannins, giving a new product that would slowly, but very noticeably, seep out and get caught on the outer, more decorative, garments. And the stains you got from that were not nice,

---

IIIIIOnce numbering in the millions, at its lowest the population of wild American bison (buffaloes) was down to less than a hundred wild animals. Today, there are around 500 000 bison in North America.

****"It was the daily companion of women from all classes and age groups, a necessary item of attire, which few 'decent' women would dare eschew."

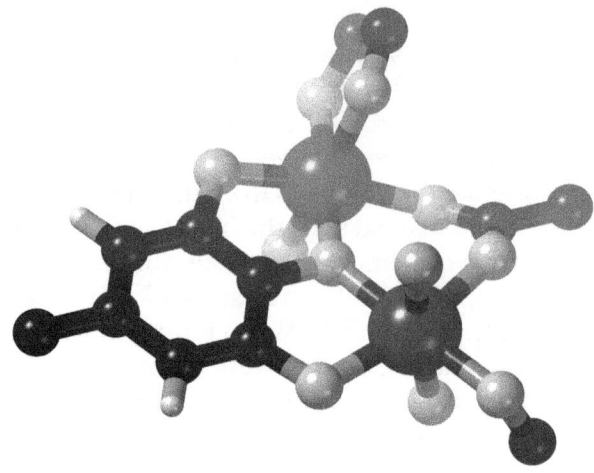

Figure 10.9   Fragment of a possible iron complex with tannic acid, the basic
unit of gall ink. Similar blueish-black compounds would form
by polyphenols leaching from vegetable tanned leather and
combining with corroded corset iron, posing a considerable
social risk for fashionable Victorian ladies.

as this new chemical very closely resembled the most popular
type of writing ink used in Europe for the last 700 years.

Known since the antiques, this was gall ink, not prepared
using bark extracts but from gallnuts, very rich in tannic acid, the
most popular ones being the Aleppo variety. Gallnuts of course
could also be used for tanning. Mixed with iron 2+ ions, this gave
a soluble product that could easily diffuse into the paper fibres
where the iron ions eventually would be oxidised to iron 3+, and
the now insoluble tannic acid complexes were permanently fixed
to the paper, or for an unfortunate Victorian lady, on the crino-
line.[††††] Quite likely, gall ink was the first man-made commercial-
ised coordination compound (Figure 10.9).

10.5   BEYOND STEAMPUNK

As fascinated as one might be with corsets and Victorian era
inspired steampunk, we will now move on to modern times.
True, chrome tanning is still the most common way of making

[††††]To demonstrate this, first make some strong tea, then dissolve (as best you can) a small
bit of an iron supplement pill in water or vinegar. Add the iron solution to the tea and
watch the ink form.

leather, but what more can be expected from coordination chemistry as the 21st century starts to mature?

A catalyst that could help us harvest the energy from the sun would be great. We could do CO_2 reduction or hydrogen gas production from water, making fuels in a sustainable way. Or nitrogen fixation, taking the nitrogen gas from the atmosphere and turning it into fertilizers in a more energy efficient way than today's iron catalysts.

On a fundamental level, this is also about understanding photosynthesis, where many different metal ions are involved, from magnesium to copper. A problem at least as old as the 1938 comedy *You Can't Take it with You*, directed by the chemical engineer Frank Capra. Here, Tony Kirby (James Stewart) confesses to his fiancée Alice Sycamore (Jean Arthur), that he is not really interested in making money, understanding and remaking photosynthesis is his passion.

A recent example of photocatalysis using the chlorophyll from plants is the preparation of the malaria drug artemisinin (Chapter 4).[9]

In all sorts of other ways, new and more efficient catalysts are needed for emerging Green Chemistry processes, especially if we want to switch our chemical feedstock from fossil-based to bio-based raw materials.

Our friends and colleagues also work on new materials and new medicines, they try to understand the very complex chemistry of metal ions in biological systems, and they devise new methods to catch criminals by inventing better crime scene investigating tools.

Just a few examples here. The first is the 2016 Nobel Laureate Sir Fraser Stoddart, who has developed a method that could potentially save Ghana from importing more than 20 000 tons of cyanides every year. Instead of using NaCN to make the linear ion $[Au(CN)_2]^-$, his group used potassium bromide to make the square planar $[AuBr_4]^-$ ion that we used to detect cocaine in Chapter 5. Not so exciting perhaps, but what followed next was a real discovery; by adding a naturally occurring sugar, alpha-cyclodextrin, to the solution, a solid was immediately formed. This solid efficiently removed all gold from the solution, giving a very pure substance composed of one large sugar molecule, one potassium ion encircled by six waters, and the gold(III) ion with four bromides around it (Figure 10.10).

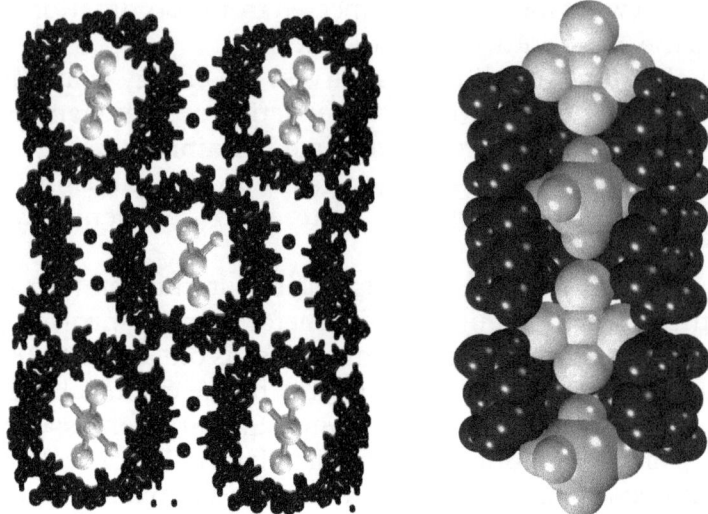

Figure 10.10   Scottish farm boy Sir Fraser Stoddart struck gold with the 2016
                   Nobel prize in chemistry, and may prove to have real impact on
                   the mining of the precious metal by the almost edible chem-
                   istry that might revolutionise how mining companies extract
                   gold. His group has replaced poisonous cyanide by cyclodex-
                   trin sugar (black) giving an exact fit of alternating $[AuBr_4]^-$ and
                   $[K(H_2O)_6]^+$ ions (white). A bigger chunk of a crystal is shown to
                   the left and a cut-through picture of the channels to the right.
                   The compound forms an easily separable solid phase that can
                   then very straightforwardly give pure gold metal.

It works because of an extraordinary match between the sizes
of the tubular cages formed by the cyclodextrins and the two ions,
and the interactions between the ions.[10] Moreover, alpha-cyclodextrin can be cheaply made from starch, and KBr has been available inexpensively on the ton scale ever since Herbert Dow started
to pump up large amounts of bromide rich salt solutions from
underground reservoirs in Ohio and Michigan in the 1890's.[‡‡‡‡]

But this is not just a laboratory experiment, tests under realis-
tic conditions are already underway at a few gold mines. These
may, by the way, be considered as holes in the ground with a

[‡‡‡‡]It is not likely that the method will be able to extract gold from seawater though, but an
     interesting coincidence is that Herbert Dow himself was one of the many who believed
     in that idea in the 1920's. The first large use of KBr was as a mild sedative, some hos-
     pitals apparently used tons of it in the 1890's and Agatha Christie later used it to con-
     struct her first crime novel.

chemical factory on top, as in fact any mine. Not that we want to diminish the hard and dangerous work of the miners. Just to remind you that it is all down to chemistry if we want any metals at all, even gold.

10.6   WATER HARVESTING IN THE DESERT

Gold is all good and fine, but as king Midas discovered, you cannot drink it. And water problems on this planet are not going to get any less pressing in the future. But even in the driest conditions there is water in the air, relative humidity seldom goes below 20 percent even in places like Botswana and Jordan. Imagine if you could turn those gaseous water molecules into liquid water with a device just powered by the hot sun?

Perhaps this will be possible. What we need is a porous material that in the cold desert night will fill its channels with water. Then, as the sun rises and heats the device, the water molecules should let go and form a stream of air oversaturated with water. This humid gas can then easily be persuaded to give away its water molecules to form liquid water.

Now, think of the molecules in the chrome-tanning picture or the malachite picture. The carboxylate ends that are attached to the chromium ions are bound to the protein in the other end, but what if they instead had another carboxylate in the other end, and a bit of a spacer in-between? Or, look at the bridging carbonate groups (CO_3^{2-}) making the malachite pigment a network, what if the third oxygen was replaced by a spacer, and the other end of the spacer was replacing an oxygen in another carbonate? Giving you something like: metal-carboxylate-spacer-carboxylate-metal.

Then you would get a new class of materials known as metal-organic frameworks, or MOFs for short. Large or small organic molecules bridging metal ions or clusters of metal ions giving network compounds with empty space in-between. Or rather, water or some other solvent molecule.

In the 1990's common wisdom, or should we be less diplomatic and say a fair number of vocal old professors, said that any material like this would collapse like a badly crafted soufflé once the solvent molecules were removed by heating or by vacuum. It turned out that they were wrong, and this is now one of the

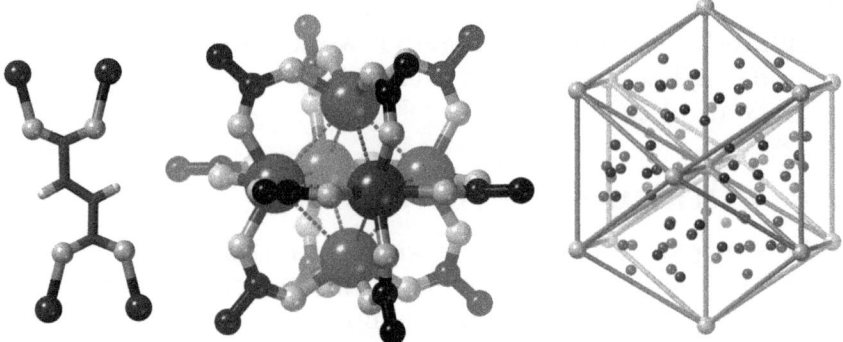

Figure 10.11   The Metal-Organic Framework in the picture (MOF-801) is used in a prototype solar-powered-desert-water-harvesting device. To the left: A fumarate carboxylate ion bridging two pairs of zirconium ions. Centre: Six zirconium ions forming an octahedron, shown with dashed lines (these are not chemical bonds). Not shown are the eight hydroxide ions binding to three zirconium ions on each face. This means twelve "rods" will stick out of the octahedron and form a network that is shown to the right, where we have replaced every octahedron with one ball and every fumarate with a rod. Water molecules are shown as black balls. This material comes out of Professor Omar Yaghi's laboratory, who was born in Amman, Jordan and describes himself as a "desert guy", who knows about the scarcity of water.

more prolific areas of inorganic chemistry. You might meet these materials in your future medicine, as they are being investigated as drug delivery materials, or in the truck hauling foodstuffs to your supermarket, because they can also be used to store natural gas (fossil origin or biogas) for combustion engines.

In Figure 10.11 you see a material that has come out of Professor Omar Yaghi's research group, and used in a prototype solar-powered-desert-water-harvesting device.[11] It is composed of small organic molecules with carboxylic acid anions in the ends (to be precise: fumaric acid anions, *trans*-$^-$OOCCH=CH-COO$^-$, also occurring in different places in nature) and a polynuclear unit with six zirconium§§§§ ions, and has the working

---

§§§§We have met the element zirconium before in this book, but just for the record: Zirconium is a transition metal, situated immediately below titanium and therefore sharing some it its properties. It is a versatile metal used for encapsulating fuels rods in nuclear power plants, and in deodorants, just to mention two examples.

name MOF-801. These six zirconium ions form an octahedron (board game aficionados recognise this as the dice with eight faces) and each edge, of which there are twelve, has a carboxylate group binding to the two zirconium ions at both ends of the edge.

The intriguing symmetry properties of these crystals, and the infinite number of ways one could chose to represent them using computer graphics, is a source of endless fascination and beauty for some. For others, these are just very practical ways to show different properties and aspects of the materials to better understand and communicate. One such property is their extremely large interior surface area, which is one requirement for a good gas sorption material, be the gas water, hydrogen or hazardous boron trifluoride, arsine, and phosphine gases, used in the semiconductor industry for microchips that go into cameras, smart phones, computers, and similar devices.

10.7   ACT LOCAL, THINK AND TALK GLOBALLY

But not all research is performed by superstars like Sir Fraser Stoddart and Omar Yaghi. In fact, most isn't, but the contributions might be equally important. It is known that to make a good narrative, journalists and university communications departments often over-emphasize the role of individuals in scientific discovery at the expense of the increasingly important group effort.[12] And the group effort is not just about working together in large research teams, is about how each team, small or large, make a contribution to the overall knowledge, and discuss results and ideas with peers, students and colleagues, in articles, or over a coffee or a beer.

So, we will give you one last example, not from a superstar, but from an emerging scholar at the beginning of her career. A research group led by Dr Banothile Makhubela at the University of Johannesburg works on both metal catalysis for biomass conversion to useful chemical feedstocks or fuel, and on finding new metal-based drugs to combat cancer, HIV, tuberculosis and malaria.

Their approach to anti-cancer compounds is to combine metal-based compounds, of a type earlier shown to be active with

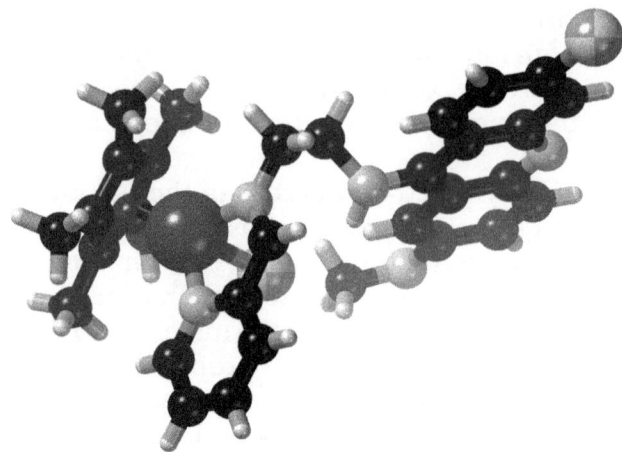

Figure 10.12   This anti-cancer rhodium compound was developed by Banothile
                   Makhubela and her group in Johannesburg, South Africa. Rh(III)
                   (large and grey) bound to a common type of cyclopentadienyl
                   ion (see "The sweet success of ferrocene" in Chapter 7), a chlo-
                   ride ion (beach ball) and a derivative of the anti-cancer drugs
                   Amsacrine and Nitracrine. This compound (the total charge
                   is +1 and a negative ion balances the charges[¶¶¶¶]) was active
                   against human leukaemia cancer cells in test tube experiments.
                   Makhubela is a recent PhD graduate from the University of Cape
                   Town (2011) and she was recognized as one of the top 200 young
                   South Africans for the year 2017 by the Mail and Guardian.[14]

existing organic compounds with known anti-cancer properties. Figure 10.12 is one example of a rhodium compound active
against human leukaemia cancer cells in test tube experiments.[13]
In contrast to the materials shown from the Yaghi and Stoddart
groups, that might be useful as they are, this is part of the continuing and difficult research conducted around the globe. A
final drug, that will have to make it through years of more test
tube experiments and clinical trials, will probably be something
different, but the knowledge gained from compounds such at
that in Figure 10.12 is a critical part of this development.

10.8   CHALLENGES AND OPPORTUNITIES

On this note we are going to end, hoping to have shown you
how metal ions and their chemistries have a vibrant history but
also that there are great challenges and opportunities ahead.

[¶¶¶¶]For the more chemically inclined of our readers we here specify that the negative ion
is tetraphenylborate, $B(C_6H_5)_4{}^-$.

Challenges more relevant than ever in our global society. And while scientists in general still have a fundamental motivation of discovering how the world works on an increasingly detailed level, in this field we are always very close to issues involving energy, environment, human health and sustainability.

Indeed, for most of us these issues are important motivations for the work we carry out, in the laboratory, supervising students and teaching graduate and undergraduate courses. It is also clear to most of us that engagement of scientists on a global level in research is essential to bring about these globally needed solutions. Sir Fraser Stoddart has noted 43 nationalities passing through his laboratory, this mixture greatly contributing to the success story of his group.

We just touched upon gold harvesting, but here at the end we also need echo the title of the first chapter. As you have seen, metal ions are indeed fantastic and we need them in everything from staying alive to make magnets for wind turbines. How to catch them is increasingly becoming an issue, and there is so much chemistry we still need to develop in order to efficiently recover and reuse these building blocks of nature, making their chemistry truly sustainable.

But science cannot be a one-way street to the richer countries of this world, we need to make science truly global, with equal opportunities to contribute and exchange information, ideas and people around the globe.[15] More than a quarter of the world's population is under 15 years of age, and in Africa more than 40%.[16] Increasingly, the pool of talent will thus be found outside the countries with the traditionally strongest research institutions.

And just as we need everybody to contribute to the global research effort, we need all branches of science to flourish and develop, not just those that seem to be the trendiest for the moment. So, we need to keep looking at the metal ions and their compounds in all sorts of ways, never mind that it all started in Zürich back in the 1890's. [IIII]

REFERENCES

1. *Kockums – The VISBY Class Corvette – Defining Stealth at Sea*, undated, http://www.loipart.com/Global/References/pdf/ VISBY_Class_Corvette.pdf.
2. A. B. Greenfield, *A Perfect Red: Empire, Espionage, and the Quest for the Color of Desire*, Harper Collins, 2009.

[IIII] With Alfred Werner, French-Swiss chemist, Nobel laureate 1913.

3. A. D. Covington, *Tanning Chemistry: The Science of Leather*, Royal Society of Chemistry, 2009.

4. M. S. Taylor, *Am. Econ. Rev.*, 2011, **101**, 3162–3195.

5. M. Odle, *"Tools & Techniques of Bark Tanning" The Book of Buckskinning VII*, Scurlock Publishing Co., Texarkana, TX, 1995, As quoted on http://www.braintan.com.

6. D. H. Ranglack and J. T. du Toit, *Oryx*, 2016, **50**, 549–554.

7. L. Summers, The sexual politics of corsetry, 1850–1900, PhD thesis, Department of History, The University of Melbourne, 1999.

8. H. Procter, *RSA J.*, 1898, **47**, 851.

9. K. Welter, *Chemistry World*, 2018, https://www.chemistryworld.com/news/antimalarial-plants-chlorophyll-catalyses-drug-synthesis/3008727.article.

10. Z. Liu, M. Frasconi, J. Lei, Z. J. Brown, Z. Zhu, D. Cao, J. Iehl, G. Liu, A. C. Fahrenbach, Y. Y. Botros, O. K. Farha, J. T. Hupp, C. A. Mirkin and J. Fraser Stoddart, *Nat. Commun.*, 2013, **4**, 1855.

11. H. Kim, S. Yang, S. R. Rao, S. Narayanan, E. A. Kapustin, H. Furukawa, A. S. Umans, O. M. Yaghi and E. N. Wang, *Science*, 2017, **356**, 430–434.

12. J. Farrar, *The Guardian*, 2017, https://www.theguardian.com/commentisfree/2017/sep/30/we-hail-individual-geniuses-success-in-science-collaboration-nobel-prize?CMP=twt_a-science_b-gdnscience.

13. A. C. Matsheku, M. Y. H. Chen, S. Jordaan, S. Prince, G. S. Smith and B. C. E. Makhubela, *Appl. Organomet. Chem.*, 2017, **31**, e3852.

14. T. Oxford, Mail & Guardian 200 Young South Africans 2017: Dr Banothile Makhubela, Mail & Guardian, 2017, http://ysa.mg.co.za/2017/dr-banothile-makhubela/.

15. L. Öhrström, P. Weiderud, M. Abu Youssef and O. M. Yaghi, *Science & Diplomacy*, 2018, 7, 2, http://www.sciencediplomacy.org/perspective/2018/global-engagement-in-science-universitys-fourth-mission.

16. *UN Demographic Yearbook 2015*, 2015, https://unstats.un.org/unsd/demographic/products/dyb/dyb2015.htm.

What Makes a Molecule a Ligand?

What then makes an organic or any other molecule bind a metal ion, and thus be a ligand? One thing is the charge, and in fact much of chemistry is governed by the opposites-attract law, meaning that the positive metal ions attract negatively charged ligands, like the chloride ions that stick to platinum 2+ in cisplatin. But what about the ammonia molecules, which are neutral?

Now we'll get a little bit technical, as we need to count to eight to get to this point. The NH_3 molecule contains the combined electrons of three neutral hydrogen atoms and one neutral nitrogen atom. If you care to go back to the periodic table you will see that hydrogen is the first element and has the atomic number 1, and thus only one electron. Nitrogen has two electrons that move so close to the nucleus that we need high-energy X-rays to budge them, a mere hydrogen atom leaves them cold. These are the *core electrons*, and as we move to higher and higher atomic number in the periodic table, they become more numerous, and tend to sit tighter, as the positive charge of the atomic nucleus increases and thus attracts them with more force.

Luckily for us, we do not need to bother with them, we only need to keep the outermost electrons in mind, what chemist call *valence electrons*. And for the atoms in the ligands that we are

The Rhubarb Connection and Other Revelations: The Everyday World of Metal Ions
By Lars Öhrström and Jacques Covès
© Lars Öhrström and Jacques Covès 2019
Published by the Royal Society of Chemistry, www.rsc.org

concerned with, you can just count the little boxes in line from the leftmost end of the table and see that for nitrogen you will end up with five valence electrons. Now combine these with one from each hydrogen and you get eight. Arrange them in three two-electrons bonds (most single bonds between atoms contain two electrons) and you are left with one pair of electrons you do not know what to do with. Well, they will stay on the nitrogen because to get a stable, low energy molecule, such as the electron count around nitrogen, and indeed all other elements from B to Ne should be eight.[†]

The simple reason Pt^{2+} reacts with ammonia is then that the ammonia *free-electron pair* will point straight out from the molecule, and even if there is no net charge on NH_3, this will be a strong enough attraction to form a stable (low energy) Pt–N bond.

Counting to eight and looking for a free-electron pair is therefore all you need to do to identify a potential ligand, and throughout this book it will be enough to count three or fewer bonds to oxygen, nitrogen, sulfur or phosphorous, and you will know that there is an extra pair of electrons on that atom to comply with the octet rule. If so, then you have a potential ligand on your hands.

[†]Sometimes this is expressed as the nitrogen "wanting" eight electrons and you may know this as the octet-rule. Obviously, nitrogen atoms do not have brains, feelings or urgings, and therefore cannot "want" anything, it is as silly as saying that a rock dropped in the ocean "wants" to go to the bottom. It is all about reaching the lowest energy state.

Bibliography

For general references to the chemical industry and chemical engineering we have used the multi-volume encyclopaedias that are the standard tools for chemical engineers and contain substantial articles on anything chemical, from nuclear energy to fragrances:

Kirk-Othmer Encyclopedia of Chemical Technology, John Wiley & Sons, New York, 1999–2012.

Ullmann's Encyclopedia of Industrial Chemistry, Wiley-VCH Verlag Gmbh & Co., Weinheim, 1999–2013.

For general references to chemistry:

F. A. Cotton and G. Wilkinson, *Advanced Inorganic Chemistry*, Wiley, New York, 1989.

N. N. Greenwood and A. Earnshaw, *Chemistry of the Elements*, Pergamon Press, Oxford, 1997.

For the history of chemistry and the chemical industry:

A. J. Ihde, *Development of Modern Chemistry*, Dover Publications, New York, 1984.

Other more general references:

J. Emsley, *Nature's Building Blocks: An A—Z Guide to the Elements*, Oxford University Press, Oxford, 2003.

J. Emsley, *Elements of Murder*, Oxford University Press, Oxford, 2005.

The Rhubarb Connection and Other Revelations: The Everyday World of Metal Ions
By Lars Öhrström and Jacques Covès
© Lars Öhrström and Jacques Covès 2019
Published by the Royal Society of Chemistry, www.rsc.org

For biological inorganic chemistry:

R. Crichton, R. Ward and R. Hider, *Metal Chelation in Medicine*, RSC, 2017.

Coordinates for most of the structural and molecular models have been obtained from the Cambridge Crystallographic Database supplied by the Cambridge Crystallographic Data Centre, 12 Union Road, Cambridge, CB2 1EZ, United Kingdom.

Subject Index